MATTER, LIFE, CONSCIOUSNESS, ETC.

MATTER, LIFE, CONSCIOUSNESS, ETC.

Second Addendum

to

Important Things

JOHN E. BEERBOWER

P.J.Bear

CONTENTS

"What was life? No one knew.
It was undoubtedly aware of itself,
so soon as it was life;
but it did not know what it was."

...

"What was life? No one knew. No one knew the actual point whence it sprang, where it kindled itself. ...If there was anything that might be said about it, it was this: ... that nothing even distantly related to it was present in the inorganic world.

...

"What then was life? It was warmth, the warmth generated by a form-preserving instability, a fever of matter, which accompanied the process of ceaseless decay and repair of albumen molecules that were too impossibly complicated, too impossibly ingenious in structure."

Thomas Mann
The Magic Mountain
(1924)

In June 2024, while working my way through Thomas Mann's seemingly interminable novel *The Magic Mountain* (a project started six weeks earlier), I encountered his multipage discourse on "life." It left me astonished.

In 1924, Mann eloquently presented the fundamental questions about life that we are confronting 100 years later: the enormity of the chasm between life and inanimate matter, the mysteries of the cell, the nature of consciousness, the source of matter, the origin of life, as well as the question "What is Life? " (For context, Erwin Schrödinger's revolutionary small book, "What is Life? The Physical Aspect of the Living Cell" appeared in 1944.)

"Three great problems that plague scientists and philosophers alike are the origins of matter, the origins of life, and the origins of mind." Sara Imari Walker, *Life as No One Knows It: The Physics of Life's Emergence* (2024), p.35.

This got me thinking that I did not do the matters justice in my book *Important Things We Don't Know* and the first addendum *Imaginings*. In addition, I had written quite a bit about these issues in my last *Wanderings* book, *Disappointments*. So, I decided to adapt and revise the relevant essays from *Disappointments*, add new material and produce a second addendum, organized around some of the more striking excerpts from Mann's dissertation.

This is it.

.

CHAPTER 1

Something (Rather Than...)

"The problem of another original procreation arose,
far more wild and mysterious than the organic:
the primeval birth of matter out of the immaterial.
In fact the abyss between material and immaterial
yawned as widely, pressed as importunately
—yes, more importunately—to be closed,
as that between organic and inorganic nature."

...

"Was that which one might call the original
procreation of matter only a disease, a growth
produced by morbid stimulation of the immaterial?"

Thomas Mann
The Magic Mountain
(1924)
(Kindle Edition, Loc.5435, 5493).

"When I chose to subtitle this book
Why There Is Something Rather Than Nothing,
I wanted to connect the remarkable discoveries
of modern science to a question that has fascinated
theologians, philosophers, ... and the general public
for more than two millennia."

Lawrence Krauss
A Universe from Nothing:
Why There Is Something Rather than Nothing
(2012) (Kindle Edition, Loc.90).

Since the beginning, philosophers have posed the question of why there is something rather than nothing. Today, we might respond "why not." We recognize that the question one finds intriguing depends upon what one views as the default position, the natural starting point. Given that "something" is necessary for the question even to be asked, why are inclined to view nothingness as the natural state and require a reason to justify somethingness? *See* Roy Sorensen, "Nothingness," *The Stanford Encyclopedia of Philosophy,* Spring 2023 Edition, Section 1.

Gottfried Wilhelm Leibniz (1646–1716) believed that logic required "sufficient reasn" for anything that exists: "the principle of sufficient reason—a principle which seems to be a metaphysical rather than a purely logical one. This states that for everything that is real there must be a sufficient reason for its being what it is–a reason sufficient to have caused it to be the way it is and no other way." Leszek Kolakowski, *Why Is There Something Rather Than Nothing?: 23 Questions from Great Philosophers* (2007), p.115. If that "reason" required its own reason, then one must look further. The resulting regress lead him to conclude that the only answer could be God, who was His own reason.

"Leibniz famously asked: why is there something rather than nothing, since nothing is easier than something? If we accept that this question is meaningful (and some philosophers deny that it is), there is only one answer: God." Kolakowski, p.117.

Curiously, the threshold question is: "what is matter?" In former days, people thought of "something" as being a thing detectable by our senses. We could see it or feel it. We still think of something as being concrete, but our understanding of concrete has evolved. *See* Sorensen, Section 4. We now believe that every something is primarily nothing, unless fields are considered to be somethings. *See Important Things We Don't Know About Nearly Everything*, pp.399-401.

"We're solid, even if space is not. Undoubtedly, none of us is anything like the void of space. Yet, surprisingly, that is only a comforting illusion. You, too, are mostly empty space. This has to do with the basic architecture of the atom." William B. Miller, *Bioverse: How the Cellular World Contains the Secrets to Life's Biggest Questions* (2023), p.22.

"By nothing, I do not mean nothing, but rather nothing—in this case, the nothingness we normally call empty space. ...For any fourth grader will tell you how much energy is contained in nothing, even if they don't know what energy is. The answer must be nothing." Krauss, p.58.

"Matter is currently defined as physical substances that occupy space and have mass. Light particles—photons—are not matter because they do not have mass. ...Matter has so far been defined by a small set of properties, such as mass, charge, and spin. The most elementary objects in our universe—the quarks and leptons of the Standard Model of Particle Physics—enter our theories as point particles. This means we do not think they have

---

any internal structure, and in fact they do not have a spatial size that we can measure."

Walker, p.131.

"Physics tells us nothing about what mass is, or what charge is: it simply tells us the range of different values that these features can take on, and it tells us their effects on other features. As far as physical theories are concerned, specific states of mass or charge might as well be pure information states: all that matters is their location within an information space."

David J. Chalmers, *The Conscious Mind: In Search of a Fundamental Theory (Philosophy of Mind)* (1996), p.302.

During the twentieth century, rather dramatic progress was made in explaining the "how" of the somethings we see: general relativity, the hot Big Bang, inflation, clustering, the creation of the heavy elements and so on, leading to The Milky Way, Earth and us. The story is well known, but the narratives of these discoveries, including my own, are quite misleading. We get an image of almost straight-line progress. The real story is much, much more complicated, even somewhat chaotic. The remarkable results can seem almost serendipitous.

Noble laureate astophysicist James Peebles provides an in depth review of the evolution of contemporary cosmology, exploring the contemporaneous disputes and controversies existing at each step along the way. P. J. E. Peebles, *The Whole Truth: A Cosmologist's Reflections on the Search for Objective Reality* (2022). He interestingly describe how some of the hypotheses seem to have gained broad acceptance before the empirical results were obtained, while others remained hotly contested until the evidence became overwhelming.

In any event, our cosmological answer seems to be that the reason we have something now, rather than nothing, is that there was a very dense point of something from which the Universe came.

As cosmology developed into a real science, it was broadly assumed that Einstein's General Theory of Relativity, shown to be consistent with a variety of empirical observations within our solar system, could be extrapolated to the far reaches of our galaxy and beyond. This assumption was certainly debateable, even if not really debated.

"When I started looking into the relativistic big bang cosmology in the early 1960s I was dismayed by the close to non-existent empirical support. In those days the status of cosmology was not even what I have learned to term a social construction, because not many in the community were confident that cosmology was a 'real science.'" Peebles, p.211.

"[T]he Newtonian gravity physics that is successfully applied to the solar system is extrapolated by some ten orders of magnitude to the length scales of Babcock's observations. **This great extrapolation could have been more commonly questioned... .**" Peebles, p.150 (emphasis added).

The problem, instead, was that it conflicted with the belief in a static, eternal Universe, an apparently unanticipated result that Einstein disliked. Einstein's solution was the introduction of the Cosmological Constant to offset the attractive force of gravity. The mounting evidence in favor of a Big Bang and the continuing expansion of the Universe appeared to moot the point.

"[T]he general theory of relativity began as an accepted social construction, non-empirically confirmed by its admirable pedigree. The theory only later graduated to a well-tested empirical construction." Peebles, pp.54-55. "The evidence we have now is that this theory gives a good description of the expansion of our universe. The idea was discussed in the 1930s, but until the 1960s the evidence for the expanding universe was meagre, the idea of cosmic expansion largely speculative." Peebles, pp.3-4.

Curiously, Krauss argues that general relativity was the product of empirical inference rather than a pure thought experiment, because Einstein was trying to explain the known (observed) discrepancy in the orbit of Mercury. Peebles concludes from the same facts that the theory's successful "prediction" of the orbit was not really confirmatory evidence since the theory was constructed with the objective of getting that result. It was not a prediction of the theory. The important evidentiary confirmations of predictions came later ("the gravitational redshift and the deflection of light by mass"). *See* Krauss, pp.1-3; Peebles, p.69.

The last 100 years have also seen huge advances in our theories of the "what" that these somethings are. And, the advances in particle physics and quantum mechanics have also complicated the story of "how." Yet, while "we have a pretty good understanding of ... the process that biased the formation of matter over antimatter in the very early universe. Yet we still do not have a concrete answer to the philosophical question, What is matter anyway?" Sara Imari Walker, *Life as No One Knows It: The Physics of Life's Emergence* (2024), p.42.

"You might naïvely think that the electron really does have a mass, a charge, and a spin, and that these are intrinsic properties of the electron. But these properties can also be considered to merely describe how electrons interact with certain measuring devices. ...We have no way to know what other properties electrons may have that we haven't measured or been able to infer by measurement of their interactions with other objects. Nor can we know if electrons still have the properties we know about when they are not being measured. ...We do not know what it is for any object to just be."

Walker, p.41.

"Elementary particle theory now is well specified and many of its predictions are definite and reliably computed. The predictions fit a wealth of well-checked reproducible tests. Could a different theory of the structure of matter fit a comparable range of empirical evidence, maybe obtained by tools other than all those that have been used to probe matter and arrive at the standard theory? The same question applies to all our natural sciences, of course. A falsification of the idea is not possible, but we can observe that if two different theories of matter passed all the tests from laboratory and particle accelerator experiments, the interpretation would require an array of coincidences that seems absurd... ."

Peebles, p.52.

What is matter actually? Anything? *See infra.*, pp.164-78.

Dark Matter

Other anomalies began to appear.

For example, various spiraling galaxies should be flying apart given their observed masses. They behaved as if they were much more massive than they appeared and as if much of that additional mass was located not in the center but around the perimeter of the disc. Also, the amount of hydrogen observed in the Universe was far too great for the quantities of neutrons and protons that observations and calculations suggested exists. And, the observable mass in the Universe seemed too little to reconcile with the most current measured estimates of the age and rate of expansion of the Universe.

Rather than suspect that there was a problem with general relativity, astrophysicists proposed that the Universe contains "dark" matter, matter that is not visible. In addition, for this greater total mass to be consistent with the observed relative smoothness of the Universe's background radiation, this new matter should not interact with radiation. *See Important Things We Don't Know*, pp.599-609 .

No one knows what dark matter is, just "something" that has the various characteristics to make the math work and reconcile our empirical data.

> "We are now virtually certain that the dark matter ... must be made of something entirely new, something that doesn't exist normally on Earth. ... But it is something! ... [D]ark matter particles are all around us—in the room in which I am typing, as well as 'out there' in space."

Krauss, p.35.

> "[T]he initial density of protons and neutrons in the universe arising out of the Big Bang, as determined by fitting to the observed abundance of hydrogen, helium, and lithium, accounts for about twice the amount of material we can see in stars and hot gas. Where are those particles? ...[W]hen we add up how much 'dark matter' has to exist to explain the motion of material in our galaxy, we find that the ratio of total matter to visible matter is not 2 to 1, but closer to 10 to 1. If this is not a mistake, then the dark matter cannot be made of protons and neutrons. There are just not enough of them."

Krauss, pp.24-25.

"[M]ost ofthe mass of the system is located between the galaxies, in a smooth, dark distribution. In fact, more than 40 times as much mass is between the galaxies as is contained in the visible matter in the system (300 times as much mass as contained in the stars alone with the rest of visible matter in hot gas around them). Dark matter is clearly not confined to galaxies, but also dominates the density of clusters of galaxies."

Krauss, p.34.

Dark Energy

Krauss describes a "What if..." proposition that he and Michael Turner prsented in 1995:

"What Michael Turner and I argued in 1995 was heretical in the extreme. **Based on little more than theoretical prejudice**, we presumed the universe was flat.

...

"Then we inferred that all available cosmological data at the time were consistent with a flat universe only if about 30 percent of the total energy resided in some form of 'dark matter,' as observations seemed to indicate existed around galaxies and clusters, but much more strangely than even this, that the remaining 70 percent of the total energy in the universe resided not in any form of matter, but rather in empty space itself."

Krauss, p.75 (emphasis added).

The collection and analyses of empirical evidence continued, leading to two important conclusion:

1. "[I]n 1998, the BOOMERANG experiment demonstrated that the universe is flat." Krauss, p.78.

"The universe hoped for by theorists seemed to be vindicated by this observation, even though it appears to conflict strongly with the estimate made by weighing clusters of galaxies. ...With this kind of exquisite data a much more precise estimate can be made of the geometry of the universe. A WMAP ... confirms to an accuracy of 1 percent that we live in a flat universe! The expectations of theorists were correct... ."

Krauss, pp.51, 52, 53.

---

2. "[T]he evidence for acceleration [became] overwhelming, almost unimpeachable." Krauss, p.84.

"[B]y early 1998, Schmidt's group published a paper demonstrating that the universe appeared to be accelerating. About six months later, Perlmutter's group announced similar results and published a paper confirming the High-Z Supernova result, in effect acknowledging their earlier error—and lending more credence to a universe dominated by the energy of empty space or, as it is now more commonly called, dark energy."

Krauss, p.82.

But, now there was a need again for something like the Cosmological Constant to offset the much, much greater mass.

"Weighing the universe by measuring the mass of galaxies and clusters yields a value a factor of 3 smaller than the amount needed to result in a flat universe. Something has to give. ...[E]ven though it was known that ten times as much matter exists in the universe as could be accounted for by protons and neutrons, even that massive amount of dark matter, comprising 30 percent of what was required to produce a flat universe, was nowhere near sufficient to account for all the energy in the universe. The direct measurement of the geometry of the universe and the consequent discovery that the universe is indeed flat meant that 70 percent of the energy of the universe was still missing, neither in nor around galaxies or even clusters of galaxies!"

Krauss, pp.54, 55.

The "answer" was the assumption of "dark energy," a something (perhaps) that must have a repulsive effect.

"'How much energy would we have to put in empty space in order to produce the observed acceleration?' the answer we come up with is remarkable. The solid curve, which fits the data best, corresponds to a flat universe, with 30 percent of the energy in matter and 70 percent in empty space." Krauss, p.86.

Moreover, the observed data also required a constant density of dark energy as the Universe expanded, so we assume/conclude that dark energy is continuously created spontaneously as needed to maintain the constant density.

"We now know the age of the universe to four significant figures. It is 13.72 billion years old! ... But now that we have it, we can confirm that there is no way that a universe with the measured expansion rate today could be this old without dark energy, and in particular, dark energy that behaves essentially like the energy represented by a cosmological constant would behave. In other words, it is energy that appears to remain constant over time."

Krauss, p.87.

And, "[t]he origin and nature of dark energy is without a doubt the biggest mystery in fundamental physics today." Krauss, p.90.

Inflation

"Inflation is the only currently viable explanation of both the homogeneity and flatness of the universe, based on what could be fundamental and calculable microscopic theories of particles and their interactions." Krauss, p.97.

"[D]uring such a rapid expansion, the region that will eventually encompass our universe will get flatter and flatter even as **the energy contained within empty space grows as the universe grows.** This phenomenon happens without the need for any hocus pocus or miraculous intervention. This is possible because **the gravitational 'pressure' associated with such energy in empty space is actually negative.** This 'negative pressure' implies that, as the universe expands, **the expansion dumps energy into space rather than vice versa.** According to this picture, when inflation ends, **the energy stored in empty space gets turned into an energy of real particles and radiation**, creating effectively the traceable beginning of our present Big Bang expansion. ... All complexities and irregularities on initially large scales ... get smoothed out and/or driven so far outside our horizon today that we will always observe an almost uniform universe... ."

Krauss, pp.150-151 (emphasis added).

I have discussed many of the issues raised by the theory of inflation in *Important Things We Don't Know*, pp.

Something from Nothing

Theorists have not been much troubled by assumption of something from nothing. Earlier, an important proposed theory tried to accommodate the observed expansion of the Universe with the assumption of a static, eternal Universe by including the continuous spontaneous creation of new matter to supply new galaxies to fill in the space left by the expansion.

"In the year that Gamow introduced the hot big bang cosmology, Herman Bondi, Tommy Gold, and Fred Hoyle introduced the steady-state cosmology [1948]... . It assumes the universe is homogeneous, apart from local fluctuations, and expanding, consistent with the observed galaxy redshifts, at a rate of increase of separation that is proportional to the separation, which is Hubble's law. Matter is assumed to be continually spontaneously created, and the new matter is expected to collect by gravity in concentrations that become galaxies. These young galaxies would replace the older ones as they moved apart, keeping the universe in a steady state. The idea of continual creation of matter was just made up, of course, but one can lodge the same complaint against the relativistic hot big bang cosmology."

Peebles, p.111.

"[A] single electron is moving along, and then at another point in space a positron-electron pair is created out of nothing, and then the positron meets the first electron and the two annihilate. Afterward, one is left with a single electron moving along. ...In the brief middle period, for at least a little while, **something has spawned out of nothing!** ...The particles that appear and disappear in timescales too short to measure are called virtual particles...

. [E]lectron-positron pairs can spontaneously appear from nothing for a bit before annihilating each other again, over any short time... ."

Krauss, pp.64,65 67 (emphassis added).

Thus, we are told, dark energy, whatever it is, is created from nothing.

Krauss reasons as follows:

"[In a closed universe,] the total positive energy, including that associated with the rest masses of particles, must be exactly compensated for by a negative gravitational energy, so that the total energy is precisely zero. So ... quantum mechanically such universes could appear spontaneously with impunity, carrying no net energy. ...[T]hese universes would be completely self-contained space-times, disconnected from our own.

...

"In order for the closed universes that might be created through such mechanisms to last for longer than infinitesimal times, something like inflation is necessary. As a result, the only long-lived universe one might expect to live in as a result of such a scenario is one that today appears flat, just as the universe in which we live appears."

Krauss, pp.167-168, 169-170.

"In one sense it is both **remarkable and exciting** to find ourselves in a universe dominated by nothing. The structures we can see, like stars and galaxies, were all created by quantum fluctuations from nothing." Krauss, p.105 (emphasis added).

Well ... .

"We may well have to accept some unexplained impressiveness as a brute fact about the universe. We can try to minimize what needs to be explained, perhaps by boiling it down to simple principles at the fundamental level. But we'll always be left with the problem of why there is something rather than nothing. We'll always be left with the problem of why the ultimate laws are the way they are. And we'll always be left with the problem of why they're so interesting."

David J. Chalmers, *Reality+: Virtual Worlds and the Problems of Philosophy* (2022), p.133.

CHAPTER 2

Life

"No one knew the actual point whence it sprang,
where it kindled itself.
...[L]ife itself seemed without antecedent.
If there was anything that might be said about it,
it was this: ... that nothing even distantly related
to it was present in the inorganic world."

Thomas Mann
The Magic Mountain
(1924)
(Kindle Edition, Loc.5292).

I.

"Have you ever wondered what makes you alive?
What makes anything alive?"

Sara Imari Walker
*Life as No One Knows It:
The Physics of Life's Emergence*
(2024), pp.1, 2.

"For generations, physicists have puzzled over life. Their theories about matter and energy have helped them understand how the universe produced galaxies and planets. But physicists have struggled to understand how lifeless chemical reactions give rise to the complexity stored in our cells." Carl Zimmer, "A Test for Life Versus Non-Life," *NYT.com*, July 31, 2024.

"[S]cientists have repeatedly demonstrated that there is nothing that separates the properties of nonliving chemistry from living chemistry: the former can easily be transformed to the latter under appropriate conditions. ...In fact, as much as we have looked, we have found **the transformation from a nonliving to a living substance is not excluded by any known law of physics or chemistry. ...What modern science has taught us is that life is not a property of matter.**"

Walker, p.6 (emphasis added).

So, life is not a property or characteristic of matter, but matter can constitute things that are alive; and our experience on Earth causes us to believe that everything that is alive is made up of matter. But, what does it mean to be alive and what is the source of life, if none of the constituent parts are alive? We believe that generally life comes only from other life, but there presumably must have been at least one exception. Similarly, we believe that cells come only from other cells, but presumably there was a first cell too, right? We are beyond physics. We must look to chemistry and biology.

As for the question what is life, we tend to formulate answers in terms of a collection of the characteristics something must have for us to consider it to be alive. (For a listing of such definitions, *see Important Things We Don't Know*, pp.535-7.) If something has all of the characteristics, then we say it is alive; if something has most of the characteristics, but not all, then we tend to rely on the "I know it when I see it" test. Such an approach makes sense if the objective is to be able to recognize life, but it falls short of telling us what life is.

Moreover, it the approach necessarily incorporates our local experiences and, so, is biased toward life like that that we know. Could alien life fail our tests? Could it, as a result, be overlooked? Obviously. ("All of our notions of the origins of life are based on a single example, so it is impossible to tell which features are inevitable and which are historical accidents." Dennis Bray, *Wetware: A Computer in Every Living Cell* (2009), p.27.) "Biologists have gotten quite good at describing life on Earth. However, this description doesn't necessarily help us understand life as a general phenomenon in our universe... ." Walker, p.8.

But, it may be that there is no alien life.

Sara Walker and her colleagues have been working on understanding the origins and essential nature of life. As described in her recent book, the fundamental premise (in my understanding) is that complexity of objects (defined by the number of discrete steps necessary to assemble the object) that exist in numbers is evidence of life. Above a certain level of complexity (or number of steps of assembly), the possibility that multiple such objects could occur accidentally becomes zero. The only conceivable sources of such complexity are Darwinian natural selection and intentional design. Only life is subject to natural selection and life can lead to consciousness which generates intention and enables design.

"[L]ife is the only thing in the universe that can make objects with many unique parts... . ...A consequence of assembly theory is that we should not expect objects with a high assembly number to ever form spontaneously." Walker, pp.95, 113.

"The key hypothesis of assembly theory is that the observation of complex objects implies that the 'information' about the steps of their formation must exist in other objects—that is, it implies a physically instantiated memory— or, if you like, a set of constraints—for their formation, which can happen only via evolution or learning."

Walker, p.90.

Reminiscent of Paley's watchmaker—the significance of evidence of design.

The group claims to have developed a method based on "assembly theory" that enables the measurement of the relevant complexity of molecules. ("[W]e might use this as a means to look for evidence of selection and evolution in measurable properties of molecules." Walker, p.95.) If the approach works, these techniques could be used to search for evidence of extraterrestrial life. The group also plans a massive automated investigation of millions of possible chemical combinations, testing the resulting molecules for signs of life based on the measured complexity (a high "assembly number"). The hope is that this research will open new insights into the nature of life.

"Dr. Walker said that she and Dr. [Lee] Cronin are working with their colleagues to extend the assembly theory of life. They also have a far more ambitious effort underway: to build what she calls 'an origin-of-life engine in the lab.' Robots will mix inert chemicals in a vast number of combinations, looking for ones that produce more complex compounds."

Carl Zimmer, "A Test for Life Versus Non-Life: In a new book, physicist Sara Walker argues that assembly theory can explain what life is, and even help scientists create new forms of it," *NYT.com*, July 31, 2024.

Walker writes:

"As the evolutionary biologists Eörs Szathmáry and John Maynard Smith have pointed out, there have been several major transitions in the evolutionary history of Earth... . These include the transition from chemistry to cellular life, prokaryotes to eukaryotes, asexual clones to sexually reproducing populations, single-celled to multicellular organisms, solitary individuals to societies, and societies to language. Each of these major evolutionary transitions is associated with new modes of information processing and storage... ."

Walker, p.235.

I have previously criticized the assertion that natural selection applies to Richard Dawkins' "memes" (roughly, ideas), arguing that the proponents are confusing natural selection with "marketplace" competition. *Important Things We Don't Know*, pp.504-8, 564-5. On that basis, I disagree with analyses that include civilization, language and technology as part of the story of Darwinian evolution., like in the quotation above.

I include among those arguments Walker's thesis that technology is (or soon will be) alive. *See infra.*, pp.19-21.

II.

DNA

"You have never been just a gene or even a set of genes.
Instead, you can safely trace your origins back
to a first, single cell within your mother's womb.
Once this first cell came into existence,
it began to do things that are not written in DNA."

Alfonso Martinez Arias
The Master Builder:
How the New Science of the Cell
Is Rewriting the Story of Life
(2023), p.15.

We have plausible theories as to how RNA emerged and how RNA could have led to DNA. Are these theories the answer?

Developmental biologist Alfonso Martinez Arias, in his recent book, *The Master Builder: How the New Science of the Cell Is Rewriting the Story of Life* (2023), argues that the science has suffered as a result of the obsession with DNA, a view echoed in other books I discuss in later sections. The obsession is attributed to the success of the books by Richard Dawkins, beginning with *The Selfish Gene* in 1976. The fact is, Martinez Arias asserts, that DNA does not create life, is not sufficient for reproduction and does not contain the instructions for the creation of any complex organism. It is, instead, effectively a "toolbox" containing the codes for the creation of hundreds of different proteins and

RNA. These "tools" are used by cells to construct the millions of species that have been present on Earth.

"To do their masterful handiwork, cells use genes, choosing which will or will not be turned on and expressed to determine when and where the products of genes are deployed. An organism is the work of cells. Genes merely provide materials for their work." *Id.*, p.14.

"DNA is just a store of information to manufacture RNA and proteins. Although DNA has a role to play in the construction of our bodies, it is cells, not genes, that make you and me what we are." *Id.*, pp.66-67.

"A number of experiments have demonstrated that triplets of bases do indeed correspond to the amino acids, with some amino acids coded by more than one triplet. There is also one triplet—ATG—that indicates where the genetic code for the amino acids making up a protein starts, and three triplets—TAA, TAG, and TGA—indicate where the code should stop being translated. This code applies to all living animals and plants; it is universal, suggesting the breathtaking possibility that all DNA descends from one successful act of molecular invention eons ago." *Id.*, p.34.

"From the perspective of a cell, the genome is the catalogue of a hardware store with a vast array of tools, fixtures, and building materials from which cells can pick and choose." *Id.*, p.81.

How do we evaluate such a claim?

Look at a list of the things we know DNA cannot do. According to Martinez Arias, the list would include the following:

1. DNA mixed with all of the necessary chemicals (amino acids, enzymes, *etc.*) will not spring to life or create a cell. *Id.*, pp.62, 156.

2. DNA is not sufficient to clone an organism. You need a living cell from the same organism in which to implant the DNA. That is why efforts to recreate extinct organisms have failed. There are no such living cells. *Id.*, pp.153-4.

"The cell that received this synthesized DNA had not been created anew. The activity of the cell may have changed, because it contained a different set of genes, but the new DNA needed to be in a cell to implement the change. Without a cell, DNA is useless. Saying that this was a new life-form was the equivalent of saying that writing a new computer program results in the physical construction of a new computer, that software can create hardware."

Id., pp.61-62.

---

3. Your (unique) fingerprints and the patterns of the irises of your eyes (also unique) are not determined by your genes. Identical twins have their own fingerprints and irises different from those of each other. *Id.*, p.21.

4. Organisms of very different bodily designs share many genes. *Id.*, pp.53-6, 66.

"[W]hy do we see the same genes across species that feature wildly different aspects and organizations?" *Id.*, p.120.

---

5. There are no genes that specify, for example, the formation of a flipper rather than an arm and hand. That instruction must come from elsewhere. *Id.*, p.132.

6. "Building animals and plants requires the creation, arrangement, differentiation, and coordinated activity of many cells and cell types. These arise after the germ cells have been set aside in a self-contained area of the organism. The germ cells have no input into how the organism is built." *Id.*, p.132.

"But we can say that the answer to many of these questions cannot lie in the genes. Throughout the process of mitosis, genes remain inert, passive molecular structures."

...

"Your eyes' photoreceptors share what they sense with other cells in the brain, cells that recognize and interpret patterns in sensations by communicating with each other. The ability to carry out these activities lies not in your DNA but in the arrangement and function of the cells that make up your eyes and your brain." *Id.*, pp.79, 66.

"{G]enes are not the only way information propagates in biology... . Our biological lineages propagate information in the substrate of bioelectric fields, in the cytosol of the cell, and in cell membranes.

...

"Both the two-tailed and two-headed worms have the same genotype, which is also the same as the wild-type worms, so the information about morphology cannot be in the genome and must be stored elsewhere."

Sara Imari Walker, *Life as No One Knows It: The Physics of Life's Emergence* (2024), pp.194, 195.

---

7. If the genes were in control and their "objective" was the multiplication of themselves, then one would expect only microbes. The multiplication of single-celled organisms would be the least costly, most efficient and fastest way to propagate genes. And, the microbes species are more durable than more complex species. *Id.*, pp.126-8.

"Single-celled organisms would be far more energy-efficient vehicles for time-traveling genes than are animals." Martinez Arias., pp.127-128.

"In cells, the replication of genes as genes is restricted. It only occurs when the volume or age of the cell leads the cell to replicate itself. When genes became components of cells, they had to abide by the terms and conditions of the cells ever afterward. Their selfishness was curtailed." *Id.*, p.128.

8. "[E]ach and every cell of an organism generally has the same DNA in it, with the same monotonous structure" (*id.*, p.14), yet .

"If all cells have the same DNA, where do they encode the shape of two hundred or more different types of cells?" *Id.*, p.78.

"If the best gene decoders were given a sequence of DNA from a previously unknown eukaryote and asked to guess what type of cell it was, they would have a hard time. Certainly they could not infer the cell's size or shape or the number and organization of its organelles. Nor could they describe how the cell moved or what it did." *Id.*, p.92.

9. Laboratory experiments show that "when planted in a fly, a human PAX6 gene will create fly eyes rather than human eyes." *Id.*, p.120.

"The variety in organization that we observe in plant, fungus, and animal forms—for example, the differences between fly eyes and human eyes, fly brains and human brains—is not found in the DNA, because, as we have seen, genes from flies and humans can be used effectively in each other." *Id.*, p.66.

Assuming these facts, one must conclude that there is something other than DNA that plays an important role in the creation and subsequent evolution of life. Martinez Arias says that that something is the cell. "What makes you and me individual human beings is not a unique set of DNA but instead a unique organization of cells and their activities." *Id.*, p.9.

"Once this first cell came into existence, it began to do things that are not written in DNA." *Id.*, p.15.

> "We are alive only so long as our cells are active—manufacturing hormones, distributing food and oxygen around the body, and conveying electricity, all of which are required to keep our hearts beating and our neurons firing. When these activities come to an end, life ends. Yet our DNA remains, and so long as it's protected from outside elements, it will continue to exist for thousands of years."

Id., pp.81-82.

III.

Origins

"[T]he origin of complex life and its further development
reduces to a single question:
How did the first competent cells come into being?"

William B. Miller
Bioverse:
How the Cellular World Contains
the Secrets to Life's Biggest Questions
(2024) p.33.

Much speculation and experimentation has been devoted to attempting to demonstrate that a cell could emerge spontaneously under naturally occurring circumstances. Some progress has been achieved with respect to membranes, RNA and amino acids, as I have previously described. *Important Things We Don"t Know: About Nearly Everything* (2022), pp.541-2. *See also*, Dennis Bray, *Wetware: A Computer in Every Living Cell* (2009), pp.144-53. But, I think it fair to say that the "spark of life" (metabolism) has remained elusive.

"Sugars and amino acids can be made and modified; large molecules such as proteins, RNA, and DNA can be synthesized from small molecule precursors. But **the only feasible way to reproduce entire processes** such as metabolism, oxidative phosphorylation, or protein synthesis in a test tube **is to use molecules already made by cells**."

Bray, *Wetware*, p.207.

"As soon as some form of hereditary material has arisen—whether DNA or something else—then the trajectory of evolution becomes unconstrained by information and unpredictable from first principles. What actually evolves will depend on the exact environment, the contingencies of history, and the ingenuity of selection."

Nick Lane, *Vital Question: Energy, Evolution, and the Origins of Complex Life (2015)*, p.23.

Biochemist Nick Lane has a theory as to how life first arose on Earth. He presents it carefully and cautiously, surrounded by caveats. His discussion is remarkably objective and even-handed. *See id.*

Science writer and commentator "Bobby" (Seyed B.) Azarian, whole heartedly embraces an older version of that theory and presents a seriously oversimplified account of it as part of, and in support of, a much grander, more ambitious claim (discussed below). *See The Romance of Reality: How the Universe Organizes Itself to Create Life, Consciousness, and Cosmic Complexity* (2024).

Azarian's description of this story of the origin of life is so articulate and clear, despite its errors, that I quote it as the introduction to my discussion here.

"The abject failure to re-create life in the lab suggested that its emergence could not have been the simple result of a warm pond getting hit by jolts of lightening, or even strong and steady sunlight... . ..[A]ttempts to manufacture a cell have ultimately been unsuccessful.

...

"How the first life-form got its energy was a great mystery until a new piece of the puzzle emerged in 1977, when eco-systems filled with new exotic life-forms were discovered deep in the ocean in conditions previously thought to be too extreme to support biology. ...[B]acteria called reductive autotrophs ... feed on the reservoirs of geothermal and geochemical energy produced by the mixing of hot magma, seawater, and rock minerals ... [and] require neither sunlight nor other organisms to sustain themselves. ...The hot and chemically diverse environment would provide not only the molecular building blocks, ... but also the high temperatures and pressures needed to organize them into a complex re-action network ... .

...

"[T]heir rocky structures contained countless pores that could function as compartments where autocatalytic sets might form.... . ...[T]he rocks provided mineral compounds and other small molecules that could act as catalysts... ."

Azarian, pp.45, 47, 48, 48-49, 50.

However, Lane's theory is that life originated in "alkaline hydro-thermal vents," not in the better known deep-sea thermal chimneys nicknamed "black smokers" where life was found in 1977.

"The discovery of submarine vents in the late 1970s came as a shock, not because their presence was unsuspected (plumes of warm water had betrayed their presence) but because nobody anticipated the brutal dynamism of 'black smokers', or the overwhelming abundance of life clinging precariously to their sides.

...

"Yet these vents, too, are misleading. They are not really cut off from the sun. The animals that live here rely on symbiotic relationships with bacteria that oxidise the hydrogen sulphide gas emanating from the smokers. ...The stunning eruption of life around these black smoker vents is therefore completely, albeit indirectly, dependent on the sun."

Lane, *Vital Question*, pp.103-4.

The submarine alkaline vent was discovered in 2000.

"[A]lkaline vents have **nothing to do with magma...** . They are **not superheated, but warm**, with temperatures of 60 to 90°C. They are not open chimneys, venting directly into the sea, but riddled with a labyrinth of interconnected micropores. And they are not acidic, but strongly alkaline. ...And the vents persist for millennia... "

Id., pp.109, 110 (emphasis added).

"Alkaline vents are not produced by the interactions of water with magma but by a much gentler process—**a chemical reaction between solid rock and water.** Rocks derived from the mantle, rich in minerals such as olivine, react with water to become the hydrated mineral serpentinite.

...

"Olivine is rich in ferrous iron and magnesium. The ferrous iron is oxidised by water to the rusty ferric oxide form. The reaction is exothermic (releasing heat), and generates a large amount of hydrogen gas, dissolved in warm alkaline fluids containing magnesium hydroxides."

Id., p.108 (emphasis added).

Lane approaches the issues from his expertise in chemistry. That perspective and knowledge permits quite interesting insights into the phenomena in question. For example, he notes that the production of organic materials (consisting of long chains of carbon atoms) requires the combination of hydrogen and carbon dioxide which would be unlikely in the presence of oxygen because the chemical attraction of oxygen and hydrogen (producing water) is too strong.

"The chances of life starting on an oxygenated planet is arguably close to zero: hydrogen must react with CO2 to form organic molecules, but does so very reluctantly if at all in the presence of oxygen, which reacts with H2 far more avidly than does CO2."

...

"Organic molecules don't usually part with single electrons (they deal almost exclusively with pairs of electrons) hence they do not react easily with oxygen. That's why we don't spontaneously combust, and why oxygen can accumulate in the air to such high levels. But of course organic material will burn if set alight with a spark. ... Respiration is a controlled form of combustion. ...It's sobering to realise that without the quantum rules that govern the predominantly two-electron chemistry of carbon, versus the one-electron behaviour of oxygen, the world that we know and love could not exist."

Lane, *Transformer* (2022), pp.158, 159.

He also explains why carbon is a compelling choice for the building blocks of organisms (the large number of potential connection points for other atoms, including other carbon atoms). So, Lane talks a lot about the behavior of electrons. (Also protons, the relative presence of which determines the pH or acidity of a solution).

This type of analysis leads him to the conclusion that the desirable environment for the emergence of life would be a slightly alkaline solution with plenty of CO_2, a supply of H_2 and a steady flow of free energy. Then, one would need some physical enclosure to allow the concentration of resulting materials and to keep out contaminants. So, "[a]lkaline hydrothermal vents provide **exactly the conditions required for the origin of life**: a high flux of carbon and energy that is physically channelled over inorganic catalysts, and constrained in a way that permits the accumulation of high concentrations of organics." *Id.*, p.110.

The Earth 4 billion years ago was largely covered in water. There was no oxygen present, but there was an abundance of CO_2.

"In terms of thermodynamics, CO2 will react with hydrogen (H2) to form methane (CH4). ...[T]he moderate temperatures and anoxic conditions in alkaline vents should have favoured the reaction of CO2 with H2 to form CH4. ...But —and this is a big but—H2 does not easily react with CO2. There is a kinetic barrier... . ...H2 and CO2 are practically indifferent to each other. To force them to react together requires an input of energy... . "

Id., pp.113-4.

"[B]acteria growing from H2 and CO2 can only grow when powered by a proton gradient across a membrane. ...The proton concentration—the acidity—is different on opposite sides of the membrane. ...[G]iven this difference in pH, it is quite easy for H2 to reduce CO2 to make formaldehyde. **The only question is: how are electrons physically transferred from H2 to CO2?** The answer is in the structure. **FeS minerals in the thin inorganic dividing walls of microporous vents conduct electrons.**"

Id., pp.117-8 (emphasis added).

"[T]he physical structure of alkaline vents—natural proton gradients across thin semiconducting walls—will (theoretically) drive the formation of organics. And then concentrate them. ... Add to this the fact that all life on earth uses (still uses!) proton gradients across membranes to drive both carbon and energy metabolism... ." *Id.*, p.120.

"The combination of high CO2, mildly acidic oceans, alkaline fluids, and thin, FeS-bearing vent walls is crucial, because it promotes chemistry that would otherwise not happen easily." Id., p.113.

Perhaps, metabolism arose in the pores (micro-vents) of an alkaline vent. It sounds plausible. But, that is still not life. What might have happened next? Lane speculates that in some cases a membrane developed, freeing the metabolizing process from its rock home, enabling it to float away. If that bubble somehow acquired the capacity to reproduce, then it could populate the seas. Now, that involves some serious as yet unexplained events.

So, maybe we have identified a path by which life could have arisen on Earth. Maybe.

"Biochemist Nick Lane ... suggests that metabolism, the array of chemical reactions essential to power life, came first. From this perspective, nucleic acids are a coda to metabolism and how everything becomes encased in a protocell and encoded in a protogenome—details to worry about later. ...[B]ut in fact there are too many possible ways in which metabolism, nucleic acids, and membranes might have come together to kick-start living systems into being. Frankly, I don't believe we'll ever know exactly what happened."

Martinez Arias, pp.86-87.

Yet, contrary to the extravagant claims by Azarian, I do not think that this theory, even if it survives empirical testing, resolves the "fine-tuning" problem (it simply adds additional preconditions, although ones that maybe are less improbable than the others); establishes that life, let alone complex life, was inevitable (especially given that the modern cell appears to have emerged only once in billions of years and that bacteria and archaea have remained largely the same during that time) or, therefore, that life must exist elsewhere in the Universe.

CHAPTER 3

Mysteries of the Cell

"He knew it was built up
out of myriads of such small organisms,
which had had their origin in a single one;
which had multiplied by recurrent division,
adapted themselves to the most varied uses a
nd functions, separated, differentiated themselves,
thrown out forms which were
the condition and result of their growth."
...
"They were units within the organic unit
of the cell they built up.
...[T]he conception of a living unit meant
by definition that it was built up
out of smaller units which were subordinate;
that is, organized with reference to a higher form."

Thomas Mann
The Magic Mountain
(1924)
(Kindle Edition, Loc. 5331, 5439).

I.

The Cell

"A life within a life. An independent living being ...
that forms a part of the whole."
...
"The life of an organism reposes in the life of a cell."

Siddhartha Mukherjee
The Song of the Cell:
An Exploration of Medicine
and the New Human
(2022), pp.xiv, 12.

The cell is the basis of life—of all life, at least as we know it. Every living thing is composed of cells. It is possible that there might be life somewhere in the Universe that exists without cells, but it is hard to imagine what it could be like.

"[S]mall, containing between one or two and up to about twenty carbon atoms, but most of them have fewer than ten carbons. Think of these as carbon 'skeletons', in which the carbon is bound to itself plus hydrogen and oxygen atoms... . These are the building blocks that make up cells, little more than a few hundred types of molecule in total."

Nick Lane, *Transformer: The Deep Chemistry of Life and Death* (2022), p. 9.

All cells derive from other cells, either through cell division ("mitosis"), in which a cell "bulks up" then divides into two identical cells, or through reproduction ("meiosis"), where two cells combine and create a new cell with genetic content coming half from each parent.

"In mitosis, ... [y]ou start, say, with forty-six (the number of chromosomes in human cells); the chromosomes duplicate (ninety-two), and then each daughter cell gets half: back to forty-six."

...

"The genesis of sperm and eggs ... require first halving the number of chromosomes, twenty-three each, and then restoring them back to forty-six upon fertilization."

Mukherjee, pp.99, 100.

"When the cells of plants and animals multiply, they do so through a process called mitosis, where the two so-called daughter cells each have similar volumes, components, and DNA to their parent. Organelles called centrioles are vital to this process."

..

"When it's time for the cell to grow and divide, the centrioles stop what they're doing, leading to the disassembly of the structures they support, move to opposite sides of the cell, and lay down a bridge of parallel microtubules across the cell. The length of these tracks, their distance from the plasma membrane, and the placement of their connection to the centrioles are dictated with the accuracy of a precision engineer."

...

"Once the chromosomes are all moved into their new positions, the spindle breaks down. As this happens, the cell is cleaved, right through its middle, with a new membrane forming to divide the two halves."

Alfonso Martinez Arias, *The Master Builder: How the New Science of the Cell Is Rewriting the Story of Life* (2023), pp.77, 78.

It is theoretically possible that the cell emerged independently on several occasions; however, if that happened, one would expect detectable structural differences to exist. In cells today, such differences have not been found, indicating that all cells existing today in animals, plants or as single-celled organisms have derived from a single original cell.

"All complex life on earth shares a common ancestor, a cell that arose from simple bacterial progenitors on just one occasion in 4 billion years."

...

"We now know that eukaryotes all share a common ancestor... . ...All plants, animals, algae, fungi and protists share a common ancestor—the eukaryotes are monophyletic. ...Any common ancestor is by definition a singular entity—not a single cell, but a single population of essentially identical cells."

Nick Lane, *Vital Question: Energy, Evolution, and the Origins of Complex Life* (2015), pp.1, 39, 40 .

"Take a yeast cell or some species of single-celled algae. These single cells, or modern cells, as biologist Nick Lane calls them, possess virtually all the features of the cells of vastly more complex organisms, including humans." Mukherjee, pp. 130-131.

Incredible.

II.

Multi-cell Organisms

"I'm not a 'one.' You're not a 'one' either.
Instead, each of us is an astounding consortium of life.
Each of us is a vast combination
of two essential living types.
We exist as a combination of our body cells and
an enormous array of cohabiting microbial life."

Alfonso Martinez Arias
The Master Builder:
How the New Science of the Cell
Is Rewriting the Story of Life
(2023), p.12

The miracle of the cell includes the fact that cells can and do get together to form multicellular organisms. I use "get together" very loosely. In fact, cells of a particular type can clump together forming larger entities. Examples are so-called Snowflake yeast and algae.

"The ability to come together in 'herds' allowed cells to cooperate in other novel ways... ." *Id.*, p.100.

Now, these clumps consist of multiple identical cells, but at a certain point, the aggregate can start acting as an entity separate from the individual cells, doing things that cannot be done by the cells individually.

"Leaflike 'organisms' with radiating structures resembling small veins (veinules) and containing multiple cells, appeared about 570 million years ago and flourished on ocean floors. Sponges agglomerated out of individual cells. Colonies of microorganisms organized themselves into novel 'beings,' heralding a new kind of existence." Siddhartha Mukherjee, *The Song of the Cell: An Exploration of Medicine and the New Human* (2022), p. 131.

"Why did we ever leave the single-celled world? Why did 'we' become 'we'—that is, multicellular organisms?" Mukherjee, p.130.

"This jump, from single cells roaming ponds and riverbanks to multicellular organisms with division of labor and function and an ability to create and shape space, was not foretold in genetic code. The moment the first multicellular organism emerged, the balance of power changed, and cells found ways to use genes for purposes that single-celled organisms had not conceived of."

Martinez Arias, p.97.

"For now, we know only that the experiments in multicellular cooperation that led to the animal kingdom bubbled up between two billion years ago, when the first eukaryotic cells emerged, and six hundred million years ago, when sponges first show up in fossil deposits. ...But then, very shortly after this moment of divergence, we start to see organisms in fossil deposits that look like certain types of marine animals living today—the beautiful, luminous ctenophores, or comb jellies, and the cnidaria, including jellyfish, sea anemones, and hydra... . With cnidaria comes the debut of Hox genes that some hail as the real hallmark of animal life."

Id., p.112.

"[A] cell might evolve along a specific lineage into a multicellular structure (something that's not inevitable but has happened independently on Earth at least twenty-five times)... ." Sara Imari Walker, *Life as No One Knows It: The Physics of Life's Emergence* (2024), p.237.

We can certainly imagine survival benefits in being an aggregate.

For example:

"...multicellularity evolved to support larger sizes and rapid movement, thereby enabling the organism to escape predation ... or to make faster, coordinated movements toward weak gradients of food. Evolution raced toward collective existence because 'organisms' could race away from being eaten—or, equally, race toward eating."

Id., p.134.

"Bacteria... have been observed cooperating in large communities, called biofilms, in order to survive harsh environments, like the thermal vent of a deep-sea volcano or the gut of an animal. But each individual bacterium still maintains the freedom to move and feed and reproduce as it pleases. It can exist on its own. ...In time, some cells living together reached the point where their progeny remained together, attached to each other after division, and successive generations ceased seeking their own way in the world. Why this happened, we don't know."

Id., p.101.

Curiously, we also find that some specialization can begin to appear among cells within the aggregation.

"In one of the most intriguing attempts, carried out at the University of Minnesota in 2014, a group of researchers led by Michael Travisano and William Ratcliff made a multicellular being evolve from a unicellular organism." *Id.*, p.132. Growing yeast cells in flasks (actually, just letting them multiply), they observed that some mother/daughter cells stuck together after the cell division. The researchers collected the cell clumps and then grew those in flasks.

They continued to repeat the process over and over. The first result was the appearance of really large aggregates. The next result was that a certain size, the aggregates would split in two. The two halves would keep growing. The third result was the aggregates came to exhibit a phenomenon in which a line of cells across the middle of the aggregate would die, facilitating the division of the aggregate.

Note that this experiment is not evolution by natural selection. It is more like selective breeding (without a male and female).

"It was an extraordinary turning point in the history of life on earth for individual cells to augment some functions and shed others, opting to depend on neighboring cells to step in to make up for their deficits in order to survive and thrive together."

Martinez Arias, p.101.

One could imagine that some random mutations in a few cells may have proven beneficial to the survival of the aggregate and became survivors, but that is not what seems to be happening. Instead, the changes that enable specialization appear to be potentials lying dormant in the individual identical cells.

If the result of evolution, then there must have been either (i) an adaptive mutation that benefited the individual cell and incidentally turned out to benefit the aggregate or (ii) at some prior time, the aggregates existed and a mutation in cell benefited the aggregate, leading to more successful aggregates which at some later point dispersed into individual cells carrying the mutation. How, then, did that mutated cell fair among its unmutated relatives? Rather complicated and speculative.

Unfortunately, I do not think that this line gets us to true multicell organisms. This example would be more relevant if reproduction occurred through the separation from the aggregate of a single cell that then commenced to "grow" into a new aggregate.

There is no doubt but that multi-cell organisms enjoy important benefits beyond size. Specialization of cells and division of labor would likely increase efficiency in biology, just as they do in human enterprises. There may even be opportunities for economies of scale. But, what's in it for the individual cell? Adaptive changes may promote the reproduction of the multicellular organism, but will the individual cell benefit? Certainly, single-celled life has not disappeared.

Now jump ahead to today's most sophisticated multicellular organisms—mammals. The complex organism arises from a single fertilized cell, which begins to divide. The descendant cells, all identical at creation, begin to specialize, forming a heart, kidneys, a liver, blood, capillaries, skin and a brain. The list goes on.

How did the potential for all these diverse functions and designs get incorporated into that single fertilized egg?

"Embryos are small structures—from 0.5 to 1 millimeter long, depending on the species—made up of thousands of cells that, in a rich display of diversity, presage the range of different cell types that will configure the organism. The number and precise organization of these different cells beg the question of how they come about from the one cell, the original zygote. The diversity is enormous.. ."

Martinez Arias, p.136.

"Each cell in the very early embryo has the potential to give rise to a whole organism." *Id.*, p.140.

So, the question of how multicellular organisms arose becomes more significant.

And,

"[P]erhaps the most astonishing feature of multicellularity is that it evolved independently, and in multiple different species, not just once, but many, many times. ...Collective existence—above isolation—was so selectively advantageous that the forces of natural selection gravitated repeatedly toward the collective."

Mukherjee, p.131.

Indeed.

But, Mukherjee also asserts: "[T]he evolution of multicellularity was not an accident, but purposeful and directional." *Id.*, p.135. Really? What is the "purpose" and from where did it come? Who or what provided the "direction"? And, while we are at it, what is the direction?

Well, the sentence has a nice ring to it, even if it makes no sense.

III.

The Modern Cell

"Few things are as inscrutable as a cell."

Nick Lane
Transformer:
The Deep Chemistry of Life and Death
(2022), p.3.

"The origin of the modern cell
is an evolutionary mystery.
It seems to have left only the scarcest
of fingerprints of its ancestry or lineage,
with no trace of a second or third cousin,
no close-enough peers that are still living,
no intermediary forms."

Siddhartha Mukherjee
The Song of the Cell:
An Exploration of Medicine and the New Human
(2022), p. 72.

Life appeared on Earth perhaps as long as 4 billion years ago, when Earth was less than a billion years old. It consisted of single-celled organisms without a nucleus. And, from where could that first cell have come? (We consider that question in a following section.) Over the next billion years, what are now called prokaryotes evolved.

"The first cells—the simplest, most primitive of our ancestors—arose on Earth some 3.5 to 4 billion years ago, about 700 million years after the birth of the Earth. ...Evolution would select more and more complex features of the cell, eventually replacing RNA with DNA as the information carrier. ...Bacteria evolved out of that simple progenitor about 3 billion years ago... ."

Mukherjee, p. 70.

"Life arose around half a billion years after the earth's formation, perhaps 4 billion years ago, but then got stuck at the bacterial level of complexity for more than 2 billion years, half the age of our planet. Indeed, bacteria have remained simple in their morphology (but not their biochemistry) throughout 4 billion years."

Nick Lane, *Vital Question: Energy, Evolution, and the Origins of Complex Life* (2015), p.1.

There were two varieties—bacteria and (what we now call) archaea. They are superficially identical, but there are structural differences suggesting independent origination. And, the archaea are more complex. These life forms persisted and thrived alone for another billion years and continue today along side of the newer eukaryotes.

"This new group might have lacked the complexity of eukaryotes, but the genes and proteins that they did have were shockingly different from those of bacteria. This ... group of simple cells became known as the archaea... .

...

But at the arcane level of their genes and biochemistry, the gulf between bacteria and archaea is as great as that between bacteria and eukaryotes (us). ...[A]rchaea have a few sophisticated molecular machines resembling those of eukaryotes, if with fewer parts—the seeds of eukaryotic complexity."

Id., pp.8, 9.

Then, about 2 billion years ago, a new type of organism appeared, called eukaryotes, comprised of cells with nuclei. They were still presumably quite simple structures.

All three life forms coexisted and thrived.

To the surprise of many biologists, genetic analyses have apparently established with relative certainty that the eukaryotes were created by the combination of an archaeon cell and a bacterial cell (a process called endosymbiosis), with the bacterial cell shedding its no longer needed genetic material to become what we call mitochondria. Subsequently, a second endosymbiosis occurred between this eukaryotes and a bacteria containing chloroplasts. This organism became capable of photosynthesis, converting the energy in sunlight into energy that could power life. Any energy constraints on life were thereby eliminated.

"Over the last few years, comparisons of large numbers of genes in more representative samples of species have come to the unequivocal conclusion that the host cell was in fact an archaeon—a cell from the domain Archaea. All archaea are prokaryotes. By definition, they don't have a nucleus or sex or any of the other traits of complex life... . In terms of its morphological complexity, the host cell must have had next to nothing. Then, somehow, it acquired the bacteria that went on to become mitochondria. Only then did it evolve all those complex traits."

...

"This radical proposition—complex life arose from a singular endosymbiosis between an archaeon host cell and the bacteria that became mitochondria—was predicted by the brilliantly intuitive and free-thinking evolutionary biologist Bill Martin, in 1998... . ...Chloroplasts are found only in algae and plants, hence were most likely acquired in an ancestor of those groups alone. That puts them as a relatively late acquisition. Mitochondria, in contrast, are found in all eukaryotes ... and so must have been an earlier acquisition. ...[S]omehow, it acquired the bacteria that went on to become mitochondria. Only then did it evolve all those complex traits."

Id., p.10.

Endosymbiosis, the successful combination of two organisms into one, is an extremely rare, arguably a"freak," occurrence. *Id.*, p.234. It may have happened only twice in 4 billion years among many, many billions of organisms. (Curiously, it has been reported in April 2024 that researchers have observed in a lab a third occurrence of Endosymbiosis, with an algae cell incorporating a bacterial cell, perhaps enabling the conversion of nitrogen in its metabolism. Of course, it is far too soon to conclude that it will be successful.)

"Endosymbiosis ... has only happened three known times. ...The first event was roughly 2.2 billion years ago ... when a single-celled organism called archaea swallowed up a bacterium that eventually became the mitochondria. ...Th[e] second event occurred when more advanced cells absorbed cyanobacteria. Cyanobacteria can harvest energy from sunlight and they eventually become organelles called chloroplasts... .

...

"[Researchers claim to have] observed primary endosymbiosis–two lifeforms merging into one organism. This incredibly rare event occurred between a type of abundant marine algae and a bacterium was observed in a lab setting. ...With this latest endosymbiosis event, it's possible that the algae is converting nitrogen from the atmosphere into ammonia that it can use for other cellular processes ... [with] the help of a bacterium. "

Laura Baisas, "For the first time in one billion years, two lifeforms truly merged into one organism," *Popular Science*, April 18, 2024.

Thereafter, a new type of eukaryotes appeared, with all the features of a modern cell. How?

Presumably, the new eurokrotes began to evolve. The host archaea had a more complex structure than did bacteria. The membrane of the bacteria may have formed the membrane of a new feature, the nucleus. That membrane could have protected the DNA of the host archaea from contamination by other components of the bacteria. The incorporated bacteria presumably gradually shed genes that were no longer useful, thereby conserving energy. The process remains a mystery because no intermediary organisms have survived. Nick Lane theorizes that the intermediaries were not sufficiently stable to last. So, we jump straight to a fully developed modern cell. (The few organisms that look like intermediaries have apparently been shown to be later in time, being modern cells that subsequently lost features not beneficial in the niches they came to occupy.)

"About 2 billion years ago (once again the exact date is a matter of debate), evolution took a strange and inexplicable turn. That is when a cell that is the common ancestor of human cells, plant cells, fungi cells, animal cells, and amoebal cells appeared on Earth." Mukherjee, p. 71.

"[M]itochondria and chloroplasts were indeed derived from bacteria by endosymbiosis, but that the other parts of complex cells probably evolved by conventional means. The question is: when, exactly? " Lane, Vital Question, pp.9-10.

It appears that, in the end, there was an organism that was to be the common ancestor of all eurokrotes existing today. That last common ancestor had essentially all of the characteristics of a modern cell and, presumably, then evolved by natural selection into all of the diverse forms of complex life the Earth has known

"We know that the common ancestor had a nucleus, where it stored its DNA. The nucleus has a great deal of complex structure that is again conserved right across eukaryotes. It is enclosed by a double membrane, or rather a series of flattened sacs that look like a double membrane but are in fact continuous with other cellular membranes. The nuclear membrane is studded with elaborate protein pores and lined by an elastic matrix; and within the nucleus, other structures such as the nucleolus are again conserved across all eukaryotes."

...

"All these genomes lead back to the last common ancestor of eukaryotes, which had more or less everything. But where did all these parts come from? The eukaryotic common ancestor might as well have jumped, fully formed, like Athena from the head of Zeus. ...How and why did the nucleus evolve? What about sex? Why do virtually all eukaryotes have two sexes? Where did the extravagant internal membranes come from? How did the cytoskeleton become so dynamic and flexible? Why does sexual cell division ('meiosis') halve chromosome numbers by first doubling them up? Why do we age, get cancer, and die? For all its ingenuity, phylogenetics can tell us little about these central questions in biology."

Lane, *Vital Question*, pp.40, 43.

Biochemists ponder the questions of what are the constraints that have kept the bacteria and archaea from becoming bigger and more complex for billions of years and what is different about eurokrotes? The answers would appear to lie in the role of mitochondria. But, I stop here. (Lane does not.) The incredible story seems to be that the cell appears to have arisen only once on Earth; the pathway to complex cells also appears to have arisen only once and evolution of that cell resulted, not in a tree of life, but in a single Last Common Ancestor.

Just freak occurrences? One after the other?

IV.

Metabolism

"And yet underneath it all,
we are barely any closer to understanding
what breathes life into these flicks of matter."

...

"Metabolism is what keeps us alive
– it is what being alive is –
the sum of the continuous transformations
of small molecules on a timescale of nanoseconds... ."

Nick Lane
Transformer:
The Deep Chemistry of Life and Death
(2022), pp.3, 8.

As I previously described in my first book, there is debate about which part of the cell came first. *Important Things We Don't Know About Nearly Everything,* p.540. One important, but still unresolved, issue is whether metabolism (the flow of energy) arose before, and led to the creation of, genetic information or genetic information enabled the emergence of metabolism. The answer to that question would have significance to theories of the appearance of life.

"These three components (a membrane, an RNA information carrier, and a duplicator) might have defined the first cell. If a self-replicating RNA system were bound by a spherical membrane, it would make more RNA copies within the confines of the sphere and grow in size by enlarging the

---

membrane." Mukherjee, *The Song of the Cell: An Exploration of Medicine and the New Human* (2022), p.70.

"First, life arose very early ... on a water world not unlike our own. Second, by 3.5 to 3.2 billion years ago, bacteria had already invented most forms of metabolism, including multiple forms of respiration and photosynthesis. For a billion years the world was a cauldron of bacteria, displaying an inventiveness of biochemistry that we can only wonder at." Nick Lane, *Vital Question: Energy, Evolution, and the Origins of Complex Life* (2015), p.29.

Nick Lane is decidedly in the priority of metabolism camp. (I do not attempt to summarize his arguments, but I do note that his discussion is exhaustive and remarkably even-handed.)

"We need the essential elements for the chemical reactions to be readily available. Since these reactions are not spontaneous, the question is what starts the cycles running?" Lane, *Transformer*, p.153.

Lane says it is the environment: "Metabolism is driven forwards not by itself, but by the environment—ultimately, by the pressure of hydrogen." *Id.*

"The glorious hypothesis that cells are animated by the continuous directed flow of simple materials and energy turned out to be true for all life." ..."[T]he biochemical pathways that produce the basic building blocks of life are indeed conserved across practically all cells." ..."[P]erhaps the most extraordinary fact is that all cells share the same basic road plan, at least for the city centre itself." *Id.* p. 17.

"We now know that respiration takes place in the mitochondria, the so-called 'powerhouses' of the cell... ." *Id.*, p.32.

Adenosine triphosphate ("ATP") is the energy storage and delivery system used in the cells of all known living organisms, an essential feature of life. It works like a rechargeable battery, discharging or releasing energy as needed by the cell and then recharging from the available energy source to be able to deliver that energy to the cell again. Cycling continuously.

"One energy carrier stands out from all the rest by virtue of its abundance and widespread use—adenosine triphosphate (ATP). This small molecule has a tail made of three phosphate groups (a phosphate being simply a small group of atoms containing one phosphorous and three oxygen atoms). The terminal phosphate in ATP requires an input of energy to be formed but can release energy when it is lost again. ...Energy available in food molecules is used to make ATP. ... Transfer of the phosphate to another molecule releases the spring of ATP but winds up the second molecule. The receiving molecule is then able to react with yet another molecule."

Dennis Bray, *Wetware: A Computer in Every Living Cell* (2009), p.57.

Three phosphate groups are linked to one another by two high-energy bonds. ATP becomes adenosine diphosphate ("ADP") when one of the three phosphate molecules breaks free. That releases energy. (Similarly, energy is released when a phosphate is removed from ADP to form adenosine monophosphate ("AMP").) ADP (and AMP) becomes ATP when one (or two) phosphate molecules are added back. When ADP becomes ATP, what was previously a low-energy molecule becomes a high-energy molecule. ADP is constantly recycled back into ATP.

What is the source from which the energy to convert ADP (or AMP) back into ATP, to enable the reattachment of the phosphates? For plants, it is the Sun; for animals, it is the food they eat.

"[O]xygenic photosynthesis. Silently humming in trees and algae and cyanobacteria, all use virtually the same machinery to fix CO2 and generate ATP. Oxygen is the waste product... . [T]hat oxygen does not come from CO2 —it comes from water, which is split apart to extract the hydrogen—2H."

...

"Literally a transducer, chlorophyll absorbs a photon of light, red light, which excites an electron. The excited electron, zapped away from its former owner, is swiftly spirited off down an electron-transport chain embedded in the membrane."

...

"[T]he power of the sun sets electrons flowing from water ... to synthesise ATP in the first photosystem, and then reduce ferredoxin in the second. By transferring electrons from water right through the two photosystems linked in series ... oxygenic photosynthesis frees life from its hydrothermal roots."

Id., pp.173-174, 174, 175.

"The process by which glucose is oxidised goes like this. First, the C6 glucose [a sugar with 6 carbons] is split into two molecules of the C3 pyruvate, each of which is broken down to acetyl CoA. These are then fed into the Krebs cycle. One complete spin of the cycle (from pyruvate) generates three molecules of CO2 and five sets of 2H, equivalent to five molecules of H2. This hydrogen is then fed to oxygen to generate energy, in the form of ATP, via cellular respiration."

Lane,*Transformer*, p. 57.

But, how does it happen?

Lane focuses on cycles, where the reactions return the chemicals to the starting conditions, ready to repeat the same reactions again. There are two different types of phenomena involved. One is called the "Krebs cycle," in which catalysts in a series of steps convert the inputted molecule into different compounds, releasing protons and electrons (essentially, splitting hydrogen atoms). The process continues in a circle. When the starting point is reached, the missing photons and electrons are replaced from another inputted source. A continuous cycle is created.

"The Krebs cycle and cellular respiration take place inside the mitochondria ... .[T]hey brought along their bacterial metabolism to the ancestor of complex (eukaryotic) cells two billion years ago. ...The Krebs cycle supplies precursors for the synthesis of amino acids, fats, sugars and more." Lane, *Vital Question*, pp.70, 71, 72.

"If, in the first few steps of a cycle beginning with citrate, two CO_2 molecules and multiple sets of 2H are stripped out, then one full turn of the cycle would need to replenish those same constituents, or it would not be a cycle at all. But the carbon, hydrogen and oxygen that have to be replenished do not need to come in the same molecular form... ." *Id.*, p.53.

The second phenomenon is the formation of a photon differential between the two sides of a membrane which permits the transfer of electrons but not protons. So, one side has electrons seeking a place to attach (an acidic solution) and the other side has photons also seeking a home (a base or alkaline solution). Then numerous tiny ports in the membrane "pump" photons across the membrane. Astonishingly, the result is a significant electrical charge across the membrane, caused by the movement not of electrons, but of photons.

"To burn the 2H derived from the Krebs cycle, a membrane was needed." *Id.*, p.63.

"[]The 2H (derived from the Krebs cycle...) are split into their component protons and electrons. The electrons are transferred to oxygen by way of the respiratory chain of carriers embedded in the membrane itself–an electrical current, insulated by the surrounding lipids. This electrical current powers the extrusion of protons across the membrane." *Id.*, 64.

"The protons accumulate outside, giving a difference in proton concentration (which is to say, pH) between the inside and outside. Critically, because protons are positively charged, their accumulation outside generates an electrical charge across the membrane, analogous to a battery. Finally, the flow of protons back through protein turbines embedded in the membrane powers the synthesis of ATP... . Only then are protons reunited with electrons on oxygen, to form water." *Id.*, 64, 65.

"The energy released by the reaction between 2H (extracted from Krebs cycle intermediates) and oxygen is transduced into an electrical charge on the membrane. This charge is awesome." *Id.*, p.68.

"The final stage of combustion turns out to be electrical in nature. A chain of electron carriers in the mitochondrial membrane allows electrons to flow from molecules of food to oxygen, generating a continuous supply of electricity. The flow of electrons pumps protons (positively charged particles, also known as hydrogen ions) out of the mitochondria, causing a positive charge to accumulate on the outside. ...The gradient thereby created is a source of stored energy—a hydrogen fuel, no less. Most important, the proton gradient drives miniature rotating protein machines embedded in the membrane. As these nanogenerators turn, they make ATP molecules. ATP then diffuses out of the mitochondrion and into the cell, where it drives a myriad of energy-requiring processes."

Bray, pp.139, 140.

The Krebs cycle can spin in reverse, using 2H and CO^2 to form organic molecules. And, so it did.

"[A] revolutionary series of papers stretching back to 1966 ... showed that in some ancient bacteria the Krebs cycle can spin in reverse–rather than stripping 2H and CO2 from food to generate energy, it uses energy to react 2H and CO2 to form organic molecules." *Id.*, p.73.

"The discovery of non-photosynthetic bacteria in deep-sea hydrothermal vents that fix CO2 by way of the reverse Krebs cycle overhauls that conclusion. These bacteria don't need the sun to make their ATP and ferredoxin–they can do it by ancient chemistry alone." *Id.*, p.110.

I have previously discussed how epigenetics, gene transfer, the microbiome and other newer discoveries have impacted Darwin's theories during the twentieth century. *Important Things We Don't Know About Nearly Everything*, pp.164-210. The role of the cell should certainly be added to the list.

CHAPTER 4

"Intelligent" Cells

"Consciousness, as exhibited by
susceptibility to stimulus, was undoubtedly,
to a certain degree, present in the lowest,
most undeveloped stages of life... .
The lowest animal forms had no nervous systems,
still less a cerebrum; yet no one would venture to
deny them the capacity for responding to stimuli."

Thomas Mann
The Magic Mountain
(1924)
(Kindle Edition, Loc.5282).

"[C]ells have powers that DNA cannot dream of.
DNA cannot send orders to cells to move right
or left within your body orto place the heart
and the liver on opposite sides of your thorax;
nor can it measure the length of your arms or
instruct the placement of your eyes symmetrically
across the midline of your face."

Alfonso Martinez Arias
The Master Builder:
How the New Science of the Cell
Is Rewriting the Story of Life
(2023), p.13.

"[E]very one of our cells is intelligent
in its limited way."

William B. Miller, Jr.
Bioverse: How the Cellular World Contains
the Secrets to Life's Biggest Questions
(2022), p.124.

So, astonishing things happen in cells that cannot be explained by
DNA. How, through what mechanisms or processes, do cells accomplish these feats?

"The arrival of plants, fungi, and animals redefined the role of genes in the organization of life on earth and, in particular, their relationship with cells. With the advent of signaling systems, evolution gifted cells with the tools to re-make their world through exchanges of information and, importantly, with the ability to control the activities of genes in space and determine the pace of the schedules that their programs generate: cells use and control genes."

Martinez Arias, p.122.

In *Bioverse* and subsequent writings, medical doctor and evolutionary biologist William B. Miller, Jr., claims that cells are self-aware and intelligent and, perhaps, even conscious. *See* Manasee Wagh, "Every Single Cell in Your Body Could Be Conscious, Scientists Say," *Popular Mechanics*, June 10, 2024. Thus, they achieve these results through individual and group decision-making and negotiation.

"Extensive research confirms that each of the cells of our body and every one of our microbial partnering cells has its individual form of intelligence. Unquestionably, that intelligence is distinctly limited compared to our intellectual gifts." Miller, *Bioverse*, p.6.

"[E]very one of our cells is intelligent in its limited way. This defining faculty encompasses all of our body cells and our cellular microbial partners. It might even be true of viruses, too, though that is very hotly debated." *Id.*, p.124.

"Sociability and collective problem-solving are a product of intelligence, not its progenitor. The progenitor of its active manifestation in animals is the intelligent, measuring cell." *Id.*, p.197.

I dismiss Miller's high-level claims as hyperbole designed to attract attention and sell books (at which it has been at least somewhat successful). Obviously, whether a cell is "intelligent" depends on what one means by intelligent. The same for "conscious" and "self-aware." In the article about Miller cited above, the author says that Miller explained that the act of "assuming that cells are conscious" can provide new avenues of discovery in medicine. Fine. It is commonplace in science to employ assumptions in a model that enable useful analysis even though the assumptions are not necessarily true.

However, Miller makes several attention-getting claims that do have potential empirical content. (By empirical content, I mean simply that the truth or falsity of the statement matters, that it makes an observable difference in the world.) However, the book consists largely of conclusory assertions ("science tells us that...;" "recent research shows that...."). The works cited in the footnotes might have relevant factual content, but most of the controversial assertions have no footnotes.

So, Miller says:

1. "Living things are not machines. They are intelligent problem-solving creatures that operate very unlike any programmed device or computer at every scope and scale." *Id.*, p.7.

As to the asserted ability to make choices among alternatives, we must ask how we can verify that an actual choice is being made, that is, that multiple alternative choices are in fact possible and that the decision-making that occurs reflects some problem solving (promotes some identifiable objective) that is not mechanically determined, that some discretion is involved. So, what are the experiments that can be conducted to demonstrate that cells are or are not intelligent? How can we ascertain whether a cell solves a problem by making a choice or a problem gets solved as a consequence of the "choice" made?

2. "Every cell can engage in the intelligent processes of receiving, assessing, and communicating information." *Id.*, p.198. "Our current limitation is that we know many of the specifics of the physical processes by which cells trade information, but we don't actually know how they understand what is being communicated." *Id.*, p.124.

"The complexity of this cooperation might lead us to wonder whether microbes are thinking. They are not, at least in the way we do ourselves, but they can demonstrate some characteristics that we have typically considered only to occur in 'higher' animals. For example, they can be social in their behaviors, showing very high levels of coordination. They communicate with meaningful signs. These qualities enable microbes to engineer highly complex biofilms; however, these functions have an explicit purpose: protect the individual participants in the biofilm through collective problem-solving solutions to environmental stresses."

Id., p.130.

"These communication mechanisms include diffusible molecules such as homoserine lactones and peptides, A-signaling amino acids, quinoline compounds, sensor kinase proteins, and gene transfers. A comprehensive range of physical processes, including mechanotransduction, vibration, biophotons, and electron transfer, is also involved. ...Our current limitation is that we know many of the specifics of the physical processes by which cells trade information, but we don't actually know how they understand what is being communicated."

Id., p.131.

"[S]ponges have access to the signaling kit that allows animal cells to talk with each other, including the proteins behind the acronyms and nicknames BMP, Notch, Nodal, Wnt, and STAT. These proteins simply don't exist in choanoflagellates or their relatives, and their main role is to mediate communication between cells. ...Between living choanoflagellates and sponges, scientists have found no closer genetic cousins—at least so far. It's difficult to imagine such a big leap in how cells organize themselves."

Martinez Arias, pp.109, 111.

It makes sense to say that cells communicate with one another, at least by transmitting and receiving signals. But, how much and what information is exchanged? We can say that cells cooperate, but is it voluntary? Or, is it more like how all of the parts of a piece of machinery "work together"? We would not typically say that the parts of a machine cooperate.

3. "[C]ells can ... group together to make collaborative [decisions]." *Id.*, p.8.

The issue seems particularly acute with respect to the relationships between cells and the bacteria and viruses constituting the microbiome. Do bacteria end up in places where there are mutually benefits to the bacteria and the cells because they they in fact prosper there or because they in some sense they recognize the beneficial potential of that location? These questions are just, in different contexts, the questions that have hovered around Darwinian evolution for decades. For example, do ecological niches create adaptations or do the adaptations create the niches? The correct answer in any situation is the one that facilitates further theorizing. Take the example of endiosymboisis, which Miller highlights. Do the two cells recognize the potential benefits of combination or does the resulting organism simply prosper by reason of the fortuitous but accidental synergies? The same with respect to the shedding of various genetic features by one component to enable more energy-efficient specialization.

4. "Cells have a 'single-minded purpose' and demonstrate faculties of prediction, planned responses, and memory." *Id.*, p.217.

Really? "Single-minded"?

Biologist Alfonso Martinez Arias, in *The Master Builder*, makes similar claims, but his book contains much more science, and he regularly points out where we do not have answers.

Martinez Arias says:

1. "Cells don't merely multiply, regulate, communicate, move, and explore; they also count, sense force and geometry, create form, and even learn." Martinez Arias, p.13.

"[C]ells learned to use the genes they had to interact with each other, to read space and use those interactions to create shapes. It is no exaggeration to call this a learning process, because a true learning process was going on. Complexes of proteins were able to read the environment of a cell and react to it, and this complex was passed on to successive generations."

Id., p.102.

"Found in all cells in the body, mTOR senses the cell's health and available nutrients and, based on this information, decides whether it should make more nutrients, divide, or die.1 In a situation of stress, mTOR is able to read the state of the cell integrating many variables, measure them against the available nutrients, and, if possible, promote biosynthetic metabolism and growth. **We still do not understand how it makes the decision... .**"

Id., p.80 (emphasis added).

---

2. "[A]t a certain point in life's history, cells invented plants and animals by gaining the capacity to use genes to cooperate and communicate with each other on a permanent basis." *Id.*, p.15.

"[T]he core of these programs is the sequence of events created by the chemistry inside the cell, but then something needs to coordinate these across cell populations and, moreover, across different cell populations within the organism. **We do not yet know how this happens, but we do know that the cell is responsible for it... .**" *Id.*, p.115 (emphasis added).

"[C]ells made use of—or, in the way evolution does, probably invented or discovered—signaling systems: BMP, Notch, Nodal, and Wnt, which arise at the same time as multicellular organisms. ...With the advent of signaling systems, evolution gifted cells with the tools to remake their world through exchanges of information and, importantly, with the ability to control the activities of genes in space and determine the pace of the schedules that their programs generate: cells use and control genes."

Id., p.122.

"[C]ells receive and integrate signals, and then, using special proteins, they seek the control regions of genes that they need to switch on or off, as each activated circuit leads to another, and another." *Id.*, p.165.

"Communicate," probably; but, "cooperate" is a bit problematic. It implies choice.

3. "Like many organs, teeth start developing in animals as part of a conversation—what biologists like to call an interaction—between two types of tissues." *Id.*, p.189.

"When the first eukaryotic cells formed from the merger of some bacteria and archaea, something shifted in how genes pursued their ambition of eternal replication. In the first place, two or more genomes had to come to an agreement between themselves: in order to survive, they had to temper their selfishness." *Id.*, p.131.

Interaction is fine. "Conversation" seems a stretch.

Both Miller and Martinez Arias regularly describe cells as engaged in negotiations, in making compromises and sacrifices and in reaching agreements. The use of anthropomorphisms can facilitate communication but will promote understanding only when the appropriate caveats are clear. Do they actually claim that cells deliberately (consciously) make deals?

Miller seems to do so, while Martinez Arias is probably better understood as as saying that cells behavior is like what one could expect if cells were actually making "deals," as if they were conscious. As previously noted, Martinez Arias freely acknowledges that we simply do not know why much of this behavior happens. I do not want my use of the word "why" to be misunderstood. We believe we understand the useful consequences or benefits of the behavior and we have identified certain details of what happens (how the results are chemically achieved), but a very big gap still exists. What enables a cell to have memory? Where or by what are "decisions" made? How are received communications understood or interpreted?

Let's look at some examples.

Miller provides a couple of descriptions of what he deems evidence of intelligence in relatively simple organisms: slime mold and termites.

In a book I have previously discussed, *Determined: A Science of Life without Free Will* (2023) (*see Imaginings,* pp.125-34), Robert Sapolsky describes slime mold and similar examples and reaches a very different conclusion: "there is no decision-making [entity] that gets information about both sites, compares the two options, picks the better one, and leads everyone to it. Instead, ... the comparison and optimal choice emerge implicitly." Sapolsky, p.161. Of course, Sapolsky argues that even humans do not have free will.

"With remarkable prescience about these ideas in 1874, the biologist Thomas Huxley wrote about the mechanistic nature of organisms, such that they 'only simulate intelligence as a bee simulates a mathematician.'" *Id.*, pp.168-9.

Termites, ants and bees

Miller says: "[t]ermites will inspect debris and find the best tool for the optimal handling and transport of nutrients. In laboratory experiments, they will even use artificial tools over natural ones if their properties are judged better for the job." Miller, p.194. But, Sapolsky asserts that organisms such as bees and ants employ a process that he calls "scout and broadcast" (Sapolsky, p.160), which mechanically achieves beneficial results that one might characterize as "problem solving." It consists of random exploration combined with a means of signaling success. He relates that bees "dance" when they find food, doing so for longer when the source appears more attractive. The longer the dance, the more likely that other bees wandering randomly will notice and join in. Evenually, most of the bees supposedly end up in the best place. The wandering ants leave a trail. The shorter trails retain the phermones lomger. The more ants that go to the same place, the stronger the trail.

Slime mold

"[S]lime molds do not have any neurons or brains. Still, they can transfer acquired memories from cell to cell. Astonishingly, slime molds can solve mazes, unerringly find the shortest route, and remember it. ...They can settle an eight-component 'traveling salesman' problem by finding the most efficient route between eight locations. ...The slime mold transferred the learned pattern to the parts that had not yet experienced the stimulus."

Miller, p.194.

Sapolsky, however, "explains" that slime mold extends "arms" along available alternative routes in search of food, then retract the unsuccessful ones and "follow" the ones that found food.

"In a slime mold, ...single-cell amoebas have joined forces by merging into a giant, cooperative single cell that oozes over surfaces in search of food, apparently an efficient food-hunting strategy... . What used to be the individual cells are interconnected by tubules that can stretch or contract, depending on the direction of oozing.

...

"Rather than sending out scouts, the entire slime mold expands to fill both corridors, reaching both food sources. And within a few hours, the slime mold retracts from the one–oat flake corridor and accumulates around the two oats. ...[W]hen food is found, tubules contract in the direction of the food, pulling the rest of the slime mold toward it. Crucially, the better the food source, the greater the contractile force generated on the tubules."

Sapolsky, pp.163, 164, 165.

There is other surprising behavior of slime mold:

> "These creatures look like plants or fungi, but they are protists, happy to forage as single cells so long as they have food around. ...In response to a lack of food, **they signal to each other** using a chemical called cAMP ... which instructs them to come together into a large aggregate. After assembling together, the cells differentiate into one of three forms: foot cells, stalk cells, or spores... . ...[F]oot cells find an anchoring spot, and stalk cells lift the spore cells up so that they can be carried away, on the wind or by a passing animal, to germinate somewhere that food is more available. ...[Then] the foot and stalk cells die. This **strategy** of generating different types of cells, with some dying while others continue living, was passed on to the first animal cells."

Martinez Arias, pp.115-6 (emphasis added).

Strategy?

SO?

Complicated issues, certainly.

Some 15 years ago, a professor at the University of Cambridge wrote a book posing the same questions I am asking and providing a lot of scientific detail as context. *See* Dennis Bray, *Wetware: A Computer in Every Living Cell* (2009). (I think that the choice of the title failed adequately to highlight the significance of the content.) In the Preface, he explained that:

> "Over the years I had acquired a ragbag of unanswered questions relating to living systems, computers, and consciousness and it was time to think them through and put them into order. So I did indeed set out, as John Steinbeck says in his Travels with Charley, 'not to instruct others but to inform myself.'"

Bray's dedication is a quotation from the Nobel Prize acceptance speech of geneticist Barbara McClintock in 1983: "A goal for the future would be to determine the extent of knowledge the cell has of itself and how it utilizes this knowledge in a 'thoughtful' manner when challenged." He asserts that during the subsequent 30 years, despite the enormous progress in biology, her questions had not been addressed. Toward the end of the book, Bray returns to McClintock:

> "In her 1983 Nobel Prize acceptance speech, the American geneticist Barbara McClintock identified a goal for future biologists: 'To determine the extent of knowledge the cell has of itself.' ...She also wanted to know how a cell uses its knowledge of the world in a thoughtful manner when challenged. A thousand chemical voices clamor for attention every instant of an amoeba's existence. How does it decide what to do?"

Id., pp.237-238.

His answer is:

> "To me, consciousness implies intelligent awareness of self and the ability to experience introspectively accessible mental states. No single-celled organism or individual cell from a plant or animal has these properties. An individual cell, in my view, is a system that possesses the basic ingredients of life but lacks sentience. **It is a robot made of biological materials.**"

Id., Preface (emphasis added).

Clear. Yet, Bray's discussion perpetuates the ambiguity contained in the vocabulary used to discuss the behaviour of single cells:

> "[W]hat do we really know about the life of these organisms? What do they **sense**? How do they **respond**? What is **important to them**?"
>
> ...
>
> "[S]ingle cells are **aware** of their surroundings. They **detect** chemical flavors, mechanical vibrations, visual stimuli, electric fields, and gravity. They **respond** by moving or by changing their shape or internal state, selectively, **in a discerning manner**. ... Almost any animate wanderer must **make decisions**. When confronted with multiple stimuli of a possibly conflicting nature, cells have to **evaluate their options and assign priorities**. They must **actively choose** one out of many possible responses."
>
> ...

"[A]ny animal large or small that pursues a freely moving existence ... has to **detect and recognize** salient features of its environment, move to a suitable niche, detect and hunt down prey, sense and avoid predators, find a mate. A flood of information enters such a cell every second of its existence through its membrane. This must be assimilated, sorted, codified. The cell **has to choose** one integrated, coherent action that ensures its survival."

Id., pp.4, 18, 25 (emphasis added).

The context and underlying science is set forth at length in Bray's book. What it seems to me to say is the following:

The operations of cells, achieving astonishing results, is analogous to the operation of modern computers.

We say that a computer "calculates," but that is a bit misleading. A computer actually only performs simple mechanical operations which, if the computer is properly programmed, result in what we call calculations or computations. There is no volitional act, nor any intention. The computer does not comprehend what it has achieved nor even that it has achieved. We now know that computers can "learn." All that is required is the capability to assess whether an outcome of various operations is better or worse than the previous outcome. The criteria for that assessment must be provided by the programmer. If it is provided, then the computer can commence making random variations in the operations, rejecting those that generate a worse result and incorporating those that generate a better result. Over time, the computer will display the illusion that it is learning. The key is an enormous number of iterations.

How does this occur in a cell? Well, it does not. It occurs in a family of cells. Random variations occur constantly, from a number of sources. With billions of cells, the number of variations is enormous. And, selection? Survival and reproduction. Over time, this type of cell appears to "learn" to respond to its environment, but not any individual cell, the family of its relatives.

The interesting thing is that all of the impressive capabilities of cells—mobility, discernment, communication, calculation—can, and presumably did, arise out of a potentially long series of often very small adaptive changes originated randomly but selected through survival and reproduction based on improvements in function or the creation of new capabilities.

"Deeper analysis reinforces the impression of blind circuitry. ...Single cells, therefore, emphatically do not have feelings or humanlike consciousness." *Id.*, p.24.

And, memory?

"If there is a fundamental unit of life, it is not an individual organism, because no individual is isolated... . ...The fundamental unit of life is not the cell, nor the individual, but the lineage of information propagating across space and time. The branching pattern at the tips of this structure is what is alive now, and it is what is constructing the future on this planet. ...A living cell is 'just' a bag of chemicals that is animated by the memory it stores and the goals it acts on."

Sara Imari Walker, *Life as No One Knows It: The Physics of Life's Emergence* (2024), pp.17, 143-144, 223.

"Every protein molecule is subtly different, carrying not only the imprint of history, shaped by evolution over millennia, but also an echo of recent events. ... [A] swimming bacterium stores a record of its environment over the past few seconds... By comparing the stimulus level in the recent past with present conditions, the bacterium learns whether its environment is getting better or worse. It then has a basis on which to decide whether to continue swimming the same way or try some other direction."

Bray, *Wetware*, pp.229, 233-234.

"These changes are cumulative: each modification adds to those that have gone before. It is as though as each organism evolves, it builds an image of the world—a description expressed not in words or in pixels but in the language of chemistry. ...The images captured by the cell include everything from recent events to the distant past... ."

Id., p.164.

Movement, sensory perception, communication?

"The [*Amoeba proteus*] cell lives on the bottoms of streams and ponds and typically crawls over the surfaces ... by extending and retracting its pseudopodia. When it encounters obstacles in its path, regions in contact with the obstacle become quiescent while those farther away move more actively. Consequently the cell flows in a new direction. ...*Paramecia* avoid obstacles by a stylized square dance in which they reverse direction, turn through a fixed angle, and again swim forward—repeating this process as necessary."

Id., pp.11, 14.

"All free-living cells, including bacteria, amoebae, and ciliates, can detect chemicals in their surrounding media. They achieve this sense of taste and smell, as we do with our sense of smell, because molecules in the outside world stick specifically to their surfaces. Signals generated by proteins in the membrane then tell the cell about possible sources of food and potentially damaging environments. ...The molecules of a fluid move continually at high speeds, colliding with each other frequently... . Their movements are due to heat energy, and their average speed is directly related to the temperature. ...The result is diffusion. ...Because of thermal diffusion, any two molecules in a cell have a good chance of meeting each other within a short space of time."

Id., pp.17, 55.

"Sensitive to the smallest changes in structure, the subtlest nuances of chemistry, weak bonds measure the compatibility of two molecules, test the goodness-of-fit. Molecules are brought together in a myriad fleeting encounters every second by the unending riot of thermal motion. ...[P]roteins are able to recognize more than one molecular partner—that is, that they have two binding sites. These two sites, moreover, seemed able to interact with each other, talk to each other from one side of the molecule to the other. ...a protein involved in one cellular process—for example, an enzyme making glucose—could be functionally linked to another process through its second site."

Id., p.58, 61.

"One method of communication between the different cells is through contact guidance. As it moves, each bacterium deposits a trail of slime made of polysaccharides that is followed by other myxobacteria. A second mechanism depends on surface contact between cells—certain molecules on the surface of one cell meet a matching receptor on a neighbor and trigger the latter to move. This "touch and go" system helps keep the cells moving together in a swarm. ...Communication over longer distances can be also achieved by the secretion of small molecules, as in the quorum sensing of bacteria."

Id., pp.171, 174.

"Molecular systems process signals so as to produce a logical output. Interacting with other systems of proteins, specialized to produce mechanical rather than chemical effects, the output signals produce physical movements. The cell moves in an informed if not intelligent way. ...But the underlying principle from the standpoint of this book is the same—they are all forms of computation. Although built from dumb molecules, protein complexes somehow operate at a higher level. They seem to have taken a step in the direction of life."

Id., pp.87, 93.

But, the question then arises:

"Like manic pathologists at an autopsy competition, we have littered our workbenches with the dissected viscera of cells. Functional parts (organelles) and molecules of all kinds are set out in display, minutely described and labeled. But **where** in this museum of parts **do we find sensation, volition, or awareness? Which insensate substances come together, and in what sequence, to produce sentient behavior?**"

Id., p.5.

But, are we sure that that is a meaningful question? One based on a legitimate distinction reflecting reality?

My reasoning above was that the functioning of cells was comparable to the operations of a computer, with natural selection providing what for the computer would be supplied by the programmer. Computers are neither sentient nor conscious. Thus, cells presumably are neither sentient nor conscious.

The questions are obvious. Is it true that computers or robots cannot be sentient or conscious? Are sentience and consciousness real, demonstrable phenomena? If so, from where do they come? How do they arise?

"In describing the software that went into [the robot] Genghis, Rodney Brooks calls it a collection of simple autonomous computational units that, 'when placed together and in the context of the physical robot, brought that inanimate object to life. It transcended the boundary between living and nonliving.' Remarking that Genghis appeared to have intentions where none were represented internally, Brooks wonders whether the intentions and motivations of real animals might similarly emerge."

Bray, *Wetware*, p.205.

These are the topics of the following chapters.

A Note on Agenda

Often scientific discoveries or advances coincide with changes in our philosophical outlook. Think of Copernicus' heliocentric model of the solar system, Paley's watchmaker, Einstein's eternal Universe (*e.g.*, "[t]he discovery that the universe is not static, but rather expanding, has profound philosophical and religious significance, because it suggested that our universe had a beginning." Lawrence Krauss, *A Universe from Nothing: Why There Is Something Rather than Nothing*, p.4). In some cases, the science may have precipitated the new vision; in other cases, the scientist's worldview may have influenced the science. From the standpoint of scientific progress, it does not much matter which came first; but as a matter of intellectual history, these relationships are of interest.

Here, the focus on the role of the cell aligns with an interest of Miller and, perhaps, Martinez Arias in emphasizing cooperation and self-sacrifice and in downplaying the importance of competition.

"[T]he central living principle: the essential element of intelligent life is that individual self-interest is best achieved through cooperation. ...It is not 'nature red in tooth and claw' as many Darwinian evolutionists insist; instead, it is nature as collaboration, cooperation, interdependence, and competitive mutualisms." *Id.*, pp.30, 260.

"[A] key difference between genes and cells: whereas genes are selfish, cells reveal themselves to be selfless." Martinez Arias, p.129.

However, altruism and cooperation are already part of modern Darwinism. *See,* e.g., *Important Things*, pp.195-9, 509-17.

Miller asserts that this scientific discovery should modify Darwinian theory. "[W]hen cells can work together toward mutual goals to combat environmental stresses, and if that work is consistently and efficiently corrected throughout time, organisms can stay stable for millions of years." *Id.*, p.95. But, we have long recognized that cooperation on one level can enhance competitiveness on another. The theory of natural selection is not challenged, but Dawkins' "selfish gene" theory appears to be inadequate.

Saposky challenges the concept of free will. Apart from scientific curiosity, he has a social justice concern that people are being subjected to criminal punishment they do not really deserve. Ultimately, he concludes that we do not have free will or, at least, not as much as we think we do. His conclusion satisfies his social justice objectives, but it leaves the science question hanging. The significant issue for science is not how much free will there is, but whether it exists at all. Has he conceded that point?

Martinez Arias is particularly focused on the implications of the scientific theories for the fields of medicine and human health. He urges the more cautious use of antibiotics and more attention to the human microbiome. He also expounds on the potentials of new treatment techniques and approaches.

CHAPTER 5

Consciousness

"Consciousness, then, was simply
a function of matter organized into life;
a function that in higher manifestations
turned upon its avatar and became
an effort to explore and explain
the phenomenon it displayed
—a hopeful-hopeless project of life
to achieve self-knowledge, nature in recoil—
and vainly, in the event,
since she cannot be resolved in knowledge,
nor life, when all is said, listen to itself."

Thomas Mann
The Magic Mountain
(1924)
(Kindle Edition, Loc.5282).

I.

" How can consciousness be just a
mechanical play of atoms and molecules?
...I may be mistaken about what I see or feel or think or wish,
but that I see, feel, think, or wish I cannot be mistaken.
...The world, myself, all life, all history, and all science
may well be images and thoughts happening in a dream.
But the dream itself is real."

Giulio Tononi
Phi
(2012), p.22.
(Emphasis added.)

I have written quite a bit, directly and indirectly, about consciousness. *See Important Things We Don't Know About Nearly Everything*, pp.433-92, 584-600; *Imaginings*, pp.133-77. Yet, there seems always to be more to say. What has already been said just seems inadequate. It is. We simply do not have answers to the fundamental questions.

The paradoxes are striking:

- If mind and body are separate and distinct, as dualism asserts, then how does the mind cause the body to do things, which believers in free will assert it certainly does. By what mechanism can the immaterial affect the material? If there is no free will, then the separate mind is an illusion or is just irrelevant.
- Subjective experience is arguably the only thing we can know is real. Everything else could be a simulation or an illusion. This was the shocking conclusion of Descartes. "*Cogito, ergo sum*, or 'I think, therefore I am,' the most famous deduction in Western thought." Christof Koch, *Then I Am Myself the World: What Consciousness Is and How to Expand It* (2024), p.73. (Even if

the simulation triggers a feeling, the feeling itself is real, not simulated.)

"[T]he act of perception alone means that the observer is real, even if what he or she is observing is not. ...You can doubt almost everything about existence, but you can't doubt that you exist. Consciousness cannot be denied or explained away." **Seyed B. (Bobby) Azarian**, *The Romance of Reality: How the Universe Organizes Itself to Create Life, Consciousness, and Cosmic Complexity* (2024), p.190.

- Moreover, "[o]ur grounds for belief in consciousness derive solely from our own experience of it." David J. Chalmers, *The Conscious Mind: In Search of a Fundamental Theory* (1996), p.101.

"Even if we knew every last detail about the physics of the universe —the configuration, causation, and evolution among all the fields and particles in the spatiotemporal manifold—that information would not lead us to postulate the existence of conscious experience." *Id.*

"Our knowledge that conscious experience exists derives primarily from our own case, with external evidence playing at best a secondary role." *Id.*, p.102.

- Yet, the subjective feelings of other people are impossible for us to know. Even the consciousness of others can only be assumed, not demonstrated. Thus, the essence of the zombie problem— we must believe that zombies (entities that are indistinguishable from humans but are not conscious) could exist because we cannot prove that someone else is not a zombie.

"Even when we know everything physical about other creatures, we do not *know* for certain that they are conscious, or what their experiences are (although we may have good reason to believe that they are)." Chalmers, p.102.

- So, consciousness must exist separately from the physical body.

"The epistemic asymmetry associated with consciousness is much more fundamental, and it tells us that no collection of facts about complex causation in physical systems adds up to a fact about consciousness." Chalmers, pp.102-103.

- If consciousness were merely a phenomenon resulting from physical processes (say, through feedback loops in neural networks), then how do we make choices? Do we not have free will, even though we know that we do?

"Given any account of the physical processes purported to underlie consciousness, there will always be a further question: Why are these processes accompanied by conscious experience? ...[T]he fact that consciousness accompanies a given physical process is a further fact, not explainable simply by telling the story about the physical facts. In a sense, the accompaniment must be taken as brute." Chalmers, pp.106, 107.

"For a phenomenon such as life, for example, the physical facts imply that certain functions will be performed, and the performance of those functions is all we need to explain in order to explain life. But no such answer will suffice for consciousness. Physical explanation is well suited to the explanation of structure and of function." *Id.*, pp. 106-107.

"Neurobiological approaches to consciousness have recently become popular. Like cognitive models, these have much to offer in explaining psychological phenomena, such as the varieties of awareness. They can also tell us something about the brain processes that are correlated with consciousness. But none of these accounts explains the correlation: we are not told why brain processes should give rise to experience at all. From the point of view of neuroscience, the correlation is simply a brute fact." *Id.*, p. 115.

"[T]he real problem with consciousness is to explain the principles in virtue of which consciousness arises from physical systems." *Id.*, p.121.

"[T]here is a simple explanation for the success of materialist accounts in various external domains. With phenomena such as learning, life, and the weather, all that needs to be explained are structures and functions. Given the causal closure of the physical, one should expect a physical account of this structure and function. But with consciousness, uniquely, we need to explain more than structures and functions, so there is little reason to expect an explanation to be similar in kind." *Id.*, p.169.

- Most of us, I suspect, think of consciousness in terms of internal dialogue conducted in language, yet our subconscious and, even, our problem solving capabilities apparently function without and independent of language (see *Imaginings*, pp.164-9). *See also*, Carl Zimmer, "Do We Need Language to Think? A group of neuroscientists argue that our words are primarily for communicating, not for reasoning," *NYT.com*, June 19, 2024.

"We conclude that although the emergence of language has unquestionably transformed human culture, language does not appear to be a prerequisite for complex thought, including symbolic thought. Instead, language is a powerful tool for the transmission of cultural knowledge; it plausibly co-evolved with our thinking and reasoning capacities, and only reflects, rather than gives rise to, the signature sophistication of human cognition."

E. Fedorenko, S.T. Piantadosi, E.A.F. Gibson, "Language is primarily a tool for communication rather than thought," *Nature* 630, 575–586, 19 June 2024.

And, so on and on.

I am aware that I have used "consciousness" broadly and imprecisely in my prior writings. I thought that satisfactory for the purposes of my comments. Here, I will be more precise.

With respect to "the hard problem of consciousness," there is a philosophical tradion, apparently started by Thomas Nagel ("What is it Like to be a Bat" (1974)) to define consciousness as the experience of feeling things. The feelings of seeing color, hearing music, experiencing loss. Chalmers, p.xii. The material brain "feels" and has "experiences." We simply cannot explain how that happens or what it is. The emergence of this phenomenon or capability is particularly puzzling because it does not seem to serve any adaptive function (the potential "why").

But, I am not a philosopher and do not parse things so finely. I find that other similarly "hard" questions include explaining the experience of self awareness or introspection, the perception of making decisions (free will) and the functioning of the unconscious. These phenomena share certain characteristics of mystery. Other mental capabilities—like memory, cognition, recognition—seem more susceptible to scientific understanding.

For me, the problem of consciousness includes:

1. Self-awareness, me versus everyone and everything else, everything not me.

"The 'self' that is the observer arises from the self-referential act of self-modeling. By introducing the idea of an internally embedded world model, the unifying theory of reality builds a bridge between matter, mind, and cosmos." Azarian, p.246.

"Self-awareness, or self-consciousness, is the subjective experience of one's own desires and emotions." Koch, p.37.

"The experience of self is as real as any other conscious experience, such as pain or pleasure. Koch, p.128.

---

2. The experience of feeling. Not mere reflexive response, but feeling. Sentience.

"The subject matter is perhaps best characterized as 'the subjective quality of experience.' When we perceive, think, and act, there is a whir of causation and information processing, but this processing does not usually go on in the dark. There is also an internal aspect; there is something it feels like to be a cognitive agent." Chalmers, p.4.

"When I introspect, I find sensations, experiences of pain and emotion, and all sorts of other accoutrements that, although accompanied by judgments, are not only judgments... ." Chalmers, p.189.

Koch writes: "Peering even deeper, with an atomic force microscope, individual macromolecules come into focus. But never pain, pleasure, or ennui." Koch, p. 74. Clever. But, actually, we can see none of the results of neural activity. All that we can ever see are the patterns of the activity. If we discover that a particular pattern is regularly associated with a specific experience, like pain in the lower right leg, we will begin to think that we "see" the feeling when we observe that pattern. In the same manner, if we found the neural signature for *ennui*, we could see the feeling. We do not know whether there are, in fact, distinctive patterns of activity for particular feelings (Koch seems to suggest that the numerous differences among people make it unlikely); but, in all events, we still have not explained the mechanism that gives rise to feelings.

3. Internal dialogue.

"Its most prominent feature is a voice inside your head, though not everyone has such a voice. ...This 'I,' with its incessant silent speech, chattering about ten times faster than were you to speak aloud, plays an increasingly dominant role as you grow into adulthood." Koch, p.37.

4. The ability to visualize or imagine alternative scenarios, future and past.

"[C]onscious creatures really can control their own destiny by selecting the future that is consistent with their long-term goals from a menu of possible options."

...

"A higher level of self-modeling, also associated with the prefrontal cortex, allows humans to simulate themselves in a seemingly infinite variety of scenarios. With a sufficiently sophisticated global workspace comes the power of imagination, which enables us to play with the mental model of the world that evolution and adaptive learning have built up in our brains. We can travel back and forward in time in our mind... ."

Azarian, pp.206, 242.

5. The perception of making choices, decisions, of agency.

"We must not confuse basic agency (being alive) with consciousness (being aware), because you can have agency without consciousness. Free will, on the other hand, requires consciousness, though you can be conscious without having free will." Azarian, p.207.

"[T]hese machines will never be sentient, no matter how intelligent they become. Furthermore, ... they will never possess what we have: the ability to deliberate over an upcoming choice and freely decide. ...Other experiences associated with self include agency ('I made the decision') and ownership ('it was my finger that pulled the trigger')." Koch, pp.20, 38.

Koch emphasizes that subjective experiences are individual and unique. I am sure that that is true, but I think that he tries to make too much of it.

"Because we only know our own idiosyncratic view of reality, we take it for granted and assume that everyone experiences the same, although many know, in an abstract way, that our experienced realities differ in ways both small and large." Koch, p.162.

"[T]here is vast genetic and developmental diversity among the eight billion people living on planet Earth, reflected in the astonishing diversity of their brains and their ways of experiencing the world. ...[W]e take reality as given and implicitly assume that everyone experiences the same, when in fact few do." Koch, pp.56, 59.

"Because we only know our own idiosyncratic view of reality, we take it for granted and assume that everyone experiences the same, although many know, in an abstract way, that our experienced realities differ in ways both small and large." Koch, p.162.

There must be very considerable commonality among human experiences to permit the emergence and evolution of societies and cultures. We may be unable ever to know what it feels like someone else, but we can have a pretty good idea. And, of course, through communication and the arts, we develop stronger and stronger understandings. But, you never know, not for sure.

"When I wonder whether other beings are conscious, I am not wondering about their abilities or their internal mechanisms, which I may know all about already; I am wondering whether there is something it is like to be them." Chalmers, pp.164-165.

II.

THE "GHOST"

Steven Pinker, in *The Blank Slate: The Modern Denial of Human Nature* (2002, 2016), makes the following claim: "The **computational theory of mind** does more than explain the existence of knowing, thinking, and trying **without invoking a ghost in the machine...** . It also explains how those processes can be intelligent—how rationality can emerge from a mindless physical process." *Id.*, p.32 (emphasis added).

"None of this is to say that the brain works like a digital computer, that artificial intelligence will ever duplicate the human mind, or that computers are conscious in the sense of having first-person subjective experience. But ... reasoning, intelligence, imagination, and creativity are forms of information processing, a well-understood physical process." *Id.*, p.33.

But, what about consciousness? Is that too just a result of information processing? Pinker stops short, with the conclusory statement quoted above, later adding "[t]hese puzzles have an infuriatingly holistic quality to them. Consciousness and free will seem to suffuse the neurobiological phenomena at every level, and cannot be pinpointed to any combination or interaction among parts." *Id.*, p.240. I find the evidence quite unclear, as I also discuss at some length in *Important Things*. I also discuss language and consciousness in the essay "The Voice of the Unconscious" (Chapter 15, *Imaginings*), the key point being that we do not actually think using language, just often translate the thoughts into language. When we learn from something we read, we do not generally remember the specific words, but only the sense or meaning. When we remember an image, it is generally not as a stored photograph (like in "the Cloud"), but as a visual conception.

Perhaps, as Pinker notes:

"Free will and subjective experience... are alien to our concept of causation and feel like a divine spark inside us. Morality and meaning seem to inhere in a reality that exists independent of our judgments. But that **separateness may be the illusion of a brain that makes it impossible for us not to think they are separate from us.**"

Id., p.424 (emphasis added)

Yet, as Pinker also says:

> "[W]e cannot doubt the existence of our minds ... because the very act of thinking presupposes that our minds exist. But we can doubt the existence of our bodies, because we can imagine ourselves to be immaterial spirits who merely dream or hallucinate that we are incarnate. **Choice, dignity, and responsibility** are gifts that set off human beings from everything else in the universe, and **seem incompatible with the idea that we are mere collections of molecules.**"

Id., p.9 (emphasis added).

And:

> "The bittersweet process of defining ourselves by our conflicts with others is not just a subject for literature but can illuminate the nature of our emotions and the content of our consciousness. In short, **without the possibility of suffering, what we would have is not harmonious bliss, but rather, no consciousness at all.**"

Id., p. 267 (emphasis added).

Is this true? I have my doubts.

Where are we? Well, time will tell. Maybe. Meanwhile, we non-neuroscientists can only wait and continue to wonder.

III.

"[T]he brain is just an amazingly complex set of interlocking dumb mechanisms causally interacting with each other. How can conscious feelings arise from mere mechanisms?"

Christof Koch
Then I Am Myself the World
(2024), p. 74.

Christof Koch, in *Then I Am Myself the World: What Consciousness Is and How to Expand It* (2024), and Seyed B. Azarian, *The Romance of Reality: How the Universe Organizes Itself to Create Life, Consciousness, and Cosmic Complexity* (2024), both trumpet the contributions of "integrated information theory," originated by Giulio Tononi, to our understanding of consciousness.

I am unable to grasp the argument.

Koch explains:

"This starting point is what makes integrated information theory (IIT) so different... . IIT starts with consciousness, not with the brain."

...

"[Integrated information theory presupposes that **things exist to the extent they have cause-effect power**. ...Something that cannot make a difference to anything or be influenced by anything is causally impotent. It can be disregarded from the point of view of existence."

...

"The phenomenal properties of an experience—its quality or how it feels—correspond one-to-one to the physical
properties of **the intrinsic cause-effect structure** unfolded
from the **underlying substrate**. ...All quality is a structure,
not a function, a process, or a computation. One implication
is that consciousness is **nonalgorithmic;** it is not (Turing)
computable."

Koch, pp.89, 92, 93, 100-101 (emphasis added).

"[T]he ... mysterious 'Eleatic Stranger' from the ancient town of Elea
in Plato's dialogue *The Sophist* ... says, 'I suggest that everything which
possesses any power of any kind, either to produce a change in anything of
any nature or to be affected even in the least degree by the slightest cause,
though it be only on one occasion, has real being.' ... [S]omething exists if
and only if it can affect things or be affected by things."

David J. Chalmers, *Reality+: Virtual Worlds and the Problems of Philosophy* (2022), p.110.

I am fine with the first couple of propositions, but I stumble over
the assertions about "structure" and "substrate." What are they and
how do they explain consciousness, which certainly seems more like a
process than a structure?

Koch continues: "The critical difference between brains and digital
computers is at the hardware level...—...where action potentials are
relayed to tens of thousands of recipient neurons ... .[T]he integrated
information of digital computers is negligible." Koch, p.20. "The
substrate of the mental is the neocortex and allied satellite structures,

sited like a crown on top of the brain, just underneath its protective skull." Koch, p.114.

The originator of the theory says this: "Each experience ... is **a unique shape** made of integrated information—a shape that is maximally ir-reducible—the **shape of understanding**. And it is **the only shape that's really real**—the most real thing there is." Giulio Tononi, *Phi: A Voyage from the Brain to the Soul* (2012), Preface (emphasis added).

Well, that still does not answer the question.

So, the key is in the unique intricate interconnections (sometimes called the "connectome") that enable an integrated response incorporating inputs from numerous different regions of the brain. Okay, but how does that astonishing integration occur? As we often find in science, insights achieved reveal phenomena even more wonderous than that with which we began the investigation.

Actually, the alleged key is in the integration of disparate pieces of information into a single coherent "picture." Like a television screen—a million pixels, none of which even hints at the image on the screen. But, we see a picture—the integrated whole. Yet, absent the viewer, is there an image or only a million pixels? Clearly, it is us watching the television that performs the integration. The same with consciousness. The interconnection of thousands of individual inputs at a single point is not sufficient. Something needs to perform the integration, the interpretation or translation, of the totality of the inputs.

We do not have the answer.

The science commentator Bobby Azarian sets out the question clearly: "Why does the global synchronous firing of certain networks of neurons create a subjective field of experience, according to Tononi's theory?"

He attempts an answer.

> "Because the brain is not just processing information, but **integrating information into a unified whole**. Just like a computer, the brain stores and processes information, but it is how that information is shared or distributed throughout the brain network that gives rise to our rich and vivid conscious experience.
>
> …
>
> "Inside the brain, there are many different things going on in distant places. Yet, somehow we perceive it all as one unified conscious experience. Thanks to integrated information theory, we are now beginning to understand how the **perceptual binding of independently processed features** happens mathematically.
>
> …
>
> "[T]he brain generates conscious experience when coordinated global activity emerges from the local electrical interactions of billions of neurons—the synchronous firing of those neurons integrates information from multiple processing streams into a single field of experience.
>
> …

"Feedback loops running from the thalamus to the cortex ... integrate information and bind features into a cohesive perceptual landscape. ...Without feedback, the brain still functions as a physiological organ controlling autonomic functions, but consciousness fades into non-existence."

Azarian, pp. 200, 231 (emphasis added).

What?

He says: "The mind can move things around in a paradox-free manner because it is not an immaterial entity as once thought; it is **a cybernetic control unit** that can steer itself toward goals that keep it in existence." Id., p.234 (emphasis added). This statement does not really help. The metaphor reflects what happens, which we already know, not how it happens, the open question.

Then: "It seems more appropriate to say that minds are an emergent property of systems that are configured to process and integrate information." Id., p.241.

Back to emergence. The idea is almost irresistible. *But see, infra.,* pp.131-42.

Koch provides an interesting discussion of fetal development of human consciousness. Subject to various caveats and qualifications, he hypothesizes that it goes like this:

- "The birth of neurons, called neurogenesis, starts around the fifth week and is largely completed by the end of the sixteenth week."

- "[N]eocortical neurons of a fetus are not properly wired up to receive any peripheral signals until about the thirtieth week.
- "Based on the way these circuits develop, peripheral pain signals can trigger reflexes but fail to ring the consciousness alarm until well into the third trimester."
- "[T]he fetus—floating in its own isolation tank, connected to the placenta that pumps blood, nutrients, and hormones into its growing body and brain, and suffused by sedation-promoting substances—is asleep."
- "Does a fetus dream while in active sleep? If so, what would a fetus, who is a *tabula rasa*, a blank slate in terms of life memories, dream of? ...My hunch is that the fetus does not dream in the way you and I dream. But it is difficult to know for certain."
- "A third-trimester fetus is unlikely to distinguish itself from the world; it is still egoless. The extent to which it has a primitive bodily awareness, such as pleasant sensations associated with warmth and nourishment through the placenta or painful ones, is at this stage impossible to ascertain. But it cannot be ruled out."
- "Things transform abruptly during the dramatic and highly stressful events attending natural, vaginal birth. The fetus wakes up and is forced from the only home it has ever known.... . A powerful surge of noradrenaline from the *locus coeruleus* deep in the brainstem ...and the cessation of sedation... ."
- "Self-awareness and silent, inner speech develop much later. Just like dreaming, these are complex cognitive processes linked to linguistic processing that take years to mature, with boys usually delayed with respect to girls."

Koch, pp. 28, 28-29, 29, 30, 31, 32, 171.

The curious thing is that when Koch gets to his integrated information theory, he expounds a view that finds causal capability or consciousness in the embryo.

"If an organism, such as the neural net of a jellyfish, satisfies the above five postulates, it feels-like-something. ...— perhaps not too dissimilar from the experiences of a third-trimester fetus... . ...Provided the system has some itsy-bitsy intrinsic causal power, it will feel-like-something."

Koch, pp.103-104.

"Indeed, it may be that every organism on the tree of life feels-like-something, is sentient, although its phenomenal content may take a primitive form unrecognizable to us." *Id.*, p.123.

But,

"[A]n eight-week-old embryo is clearly alive but not conscious." *Id.*, p.171.

Azarian cites this conclusion as the fatal defect in the theory as an explanation of consciousness—it is missing a final step. "Taken to its logical conclusions, integrated information theory says that any system that has any amount of *phi* at all should possess some degree of conscious experience and should have some degree of causal power." Azarian, p.202.

Consider "the causal powers of virtual objects. On the face of it, virtual objects can affect one another. A virtual bat can hit a virtual ball." Chalmers, *Reality+*, p.196.

Azarian asserts that that final step is the introduction of self reference. The mental models of reality begin to incorporate the self.

"According to the cognitive philosopher Douglas Hofstadter, for a point of view to emerge, the agent's world model must model itself. We call this example of self-reference self-modeling, and for reasons that will soon be explained, it is an ability that would seem to require a brain. ...So, it seems that being conscious is not enough for what we call 'free will'—that comes later. The conscious mind only gains causal power when there is an additional level of self-modeling."

Azarian, pp.218, 221.

"What the loops do is conjure up a mental simulation of that world model, allowing the agent to run simulations of possible futures, so that the agent can select the path that it predicts is most consistent with its long-term goals. The global workspace, then, would be associated with a higher level of control, which grants new freedoms to the system, such that it can truly choose its fate."

Id., pp.236-237.

"The 'self' that is the observer arises from the self-referential act of self-modeling." *Id.*, p.246.

It certainly seems that a theory of consciousness needs to include the phenomenon of self awareness, but I doubt that we get there merely by introducing the "self "as another character in the model. Indeed, one of the striking features of consciousness, in contrast to dreams, is the unique and central role played by the self.

IV.

I previously discussed a hypothetical proposed by Australian philosopher Frank Jackson in the 1980s and compared the conclusions reached by Sean M. Carroll, *The Big Picture: On the Origins of Life,*

Meaning, and the Universe Itself (2016), p.353, and Nicholas Humphrey, *Sentience: The Invention of Consciousness* (2023), pp.96-8. *See Imaginings*, pp.177-9.

The hypothetical:

"Mary is a neuroscientist who knows everything there is to know about the color red, but she has never seen anything red. One day, Mary sees red.

"Does she learn anything new?"

I criticize the views expressed by each of the two scientists. Then, I conclude: "We should just admit that Mary had a new experience and learned something. The question to address is what are the implications?"

Not much of an answer.

I find that I actually do have an opinion about the hypothetical, because I have strong feelings about color. When I could no longer ride on the stairlift and had become confined to my power wheelchair, I had to move completely to ground floor, already equipped with ramps. So, my ground floor library, with the construction of a "roll-in" shower nearby, became my sleeping room. It is a deep burgundy in color. In and around the hundreds of books, I have had hung some 15 oil paintings, mainly of clowns and still lifes, all full of color. I love the richness, the vividness and, during the day, the play of sunlight coming through the windows. I find that the colors produce in me a sense of calm, a feeling of security. Examing the paintings brings cheerfulness. My world becomes larger. I am happy.

How does it happen? By association? The triggering of memories? The release of hormones? Something else?

I know that the colors are the consequences of differing frequencies of light, that there is nothing "there," at least nothing that resembles what I perceive. At some level, I find that virtually inconceivable My reality and that of the physicists hardly overlap. But, my world requires a conscious observer, one who experiences. Indeed, my world is that of conscious observation, the assimilation of sensory data and the output of my conscious and unconscious mind.

Yes, Mary learns something (unless you artificially define "learn" to apply only to certain types of knowledge). Mary learns what red feels like.

"But she does not know what it is like to see red. No amount of reasoning from the physical facts alone will give her this knowledge. ...[K]nowledge of what red is like is factual knowledge that is not entailed a priori by knowledge of the physical facts. The only way that the conclusion can be evaded is to deny that knowing what red experience is like gives knowledge of a fact at all. ...[I]i seems clear that the knowledge she is gaining is knowledge of a fact."

David J. Chalmers, *The Conscious Mind: In Search of a Fundamental Theory* (1996), pp.103, 104.

""[E]ven if we understood completely the mechanisms for distinguishing among the colors, down to the most niggling detail, we would still fail to understand why, alongside the nerve cells in the visual parts of the brain, which slavishly perform their task—the task of distinguishing among colors—...there should also exist a conscious experience of color—an I who sees in front of himself, vividly, a red apple, or a blue sky. No matter how I program the machine, what steps I force it to follow, I cannot see how it could see the same way I see,' said Alturi. 'It could perhaps behave like me, but I just cannot see how it would experience anything at all.'"

Giulio Tononi, *Phi: A Voyage from the Brain to the Soul* (2012), p.122.

Curiously, the mechanisms through which Mary and I perceive the colors are well understood. I quote a lengthy passage from a former theoretical physicist turned writer detailing what happens.

> "Each particle of light ends its journey in the eye upon meeting a retinene molecule, consisting of 20 carbon atoms, 28 hydrogen atoms, and 1 oxygen atom. ...In its dormant condition, each retinene molecule is attached to a protein molecule and has a twist between the eleventh and fifteenth carbon atoms. But when light strikes it ... the molecule straightens out and separates from its protein.

> ...

> "Triggered by the dance of the retinene molecules, the nerve cells, or neurons, respond. ...This change in flow of electrically charged atoms produces a change in voltage... . After a distance of a fraction of an inch, the electrical signal reaches the end of the neuron, altering the release of specific molecules, which migrate a distance of a hundred-thousandth of an inch until they reach the next neuron... ."

> ...

"In another few thousandths of a second, the electrical signals reach the ganglion neurons, which bunch together in the optic nerve at the back of the eye and carry their data to the brain. Here, the impulses race to the primary visual cortex, a highly folded layer of tissue about a tenth of an inch thick and two square inches in area, containing one hundred million neurons in half a dozen layers. The fourth layer receives the input first, does a preliminary analysis, and transfers the information to neurons in other layers."

...

"At every stage, each neuron receives signals from a thousand other neurons, combines the signals—some of which cancel one another out—and dispatches the computed result to a thousand-odd other neurons."

Alan Lightman, *Probable Impossibilities: Musings on Beginnings and Endings* (2021), pp.87, 88.

Pretty incredible.

But, nothing in this long description begins to tell us how or why the colors evoke deep emotional responses in me. Lightman admits as much. His description quoted above is of a man who sees an attractive woman. He concludes:

"What is not known is why, after about a minute, the man walks over to the woman and smiles."

Id., p.90.

"If a materialist is to hold on to materialism, she really needs to deny that Mary makes any discovery about the world at all. ...Unlike the previous options, this strategy does not suffer from internal problems. Its main problem is that it is deeply implausible. No doubt Mary does gain some abilities when she first experiences red, as she gains some abilities when she learns to ride a bicycle. But it certainly seems that she learns something else: For all she knew before, the experience of red things might have been like this, or it might have been like that, or it might even have been like nothing at all. But now she knows that it is like this."

Chalmers, pp.144-145.

"(1) Is the computer experiencing anything at all when it looks at roses?; (2) If it is, is it experiencing the same sensory color quality that we have when we look at a rose, or some quite different quality? These are entirely meaningful questions, and knowing all the physical facts does not force one answer rather than another onto us. The physical facts therefore do not logically entail the facts about conscious experience."

Chalmers, p.104.

"Explaining color to achromats [persons who cannot see color] is impossible, for colors are more than a semantic label attached to surfaces. It feels-like-something to see the colors of a fluttering flag or the setting sun. Colors, like any other experience, have what is called a quale (plural qualia), a unique feeling that makes seeing orange quite different from seeing purple and radically different from smelling garlic or touching a wet towel."

Christof Koch, *Then I Am Myself the World: What Consciousness Is and How to Expand It* (2024), p.57.

These comments have been addressed to sight, to seeing colors; but, they apply to all of the senses. You could know everything about sound, but you will experience music. The same for taste and smell. Perhaps, touch or feel is an outlier.

I need to think more about that.

V.

"Could a machine be conscious? Could an appropriately programmed computer truly possess a mind? These questions have been the subject of an enormous amount of debate over the last few decades. The field of artificial intelligence (or AI) is devoted in large part to the goal of reproducing mentality in computational machines."

David J. Chalmers, *The Conscious Mind: In Search of a Fundamental Theory* (1996), p.313.

"According to these objections, there are certain functional capacities that humans have that no computer could ever have. For example, sometimes it is argued that because these systems follow rules, they could not exhibit the creative or flexible behavior that humans exhibit... . Others have argued that computers could never duplicate human mathematical insight, as computational systems are limited by Gödel's theorem in a way that humans are not... . ...[I]t is suggested that they would have no inner life: no conscious experience, no true understanding. At best, a computer might provide a simulation of mentality, not a replication."

Id., pp. 313, 314.

Chalmers conclusion about the potential consciousness of technology is based on a "principle of invariance" he advances, that identical functional organization will result in identical functional properties, including consciousness, based upon the following assumption:

> "A natural suggestion is that consciousness arises in virtue of the functional organization of the brain. On this view, the chemical and indeed the quantum substrate of the brain is irrelevant to the production of consciousness. What counts is the brain's abstract causal organization, an organization that might be realized in many different physical substrates."

Id., p.247.

"...I will argue for a principle of organizational invariance, holding that given any system that has conscious experiences, then any system that has the same fine-grained functional organization will have qualitatively identical experiences." *Id.*, pp. 248-249.

"[T]he principle of organizational invariance, which tells us that for any system with conscious experiences, a system with the same fine-grained functional organization will have qualitatively identical conscious experiences." *Id.*, p.321.

"[A] property is an organizational invariant when it depends only on the functional organization of the underlying system, and not on any other details. ...But phenomenal properties are different. As I have argued ..., these properties are organizational invariants." *Id.*, p. 328.

Chalmers utilizes various arguments to support his conclusion. Perhaps, the most direct is as follows: Suppose that we developed an electric circuit—a silicon chips—that could perfectly perform the functions of a neuron and began replacing the neurons in a conscious brain one by one. It seems unlikely that the replacement of one neuron (or of a few) would eliminate consciousness. Will there come a point at which the brain suddenly loses consciousness or will a fully artificial brain still be conscious?

Chalmers asserts that the most likely result is a conscious brain consisting entirely of these man-made circuits.

"[I]n an ordinary computer that implements a neuron-by-neuron simulation of my brain, there will be real causation going on between voltages in various circuits, precisely mirroring patterns of causation between the neurons. For each neuron, there will be a memory location that represents the neuron, and each of these locations will be physically realized in a voltage at some physical location. It is the causal patterns among these circuits, just as it is the causal patterns among the neurons in the brain, that are responsible for any conscious experience that arises."

Id., p.321.

Of course, before even getting to the hypothetical, we can challenge the basic assumption quoted above. Until we have a satisfactory explanation of consciousness, we are simply speculating about the nature of its relationship with the physical brain.

"It is certainly true that for many properties, simulation is not replication. Simulated heat is not real heat. On the other hand, for some properties, simulation is replication. For example, a simulation of a system with a causal loop is a system with a causal loop." *Id.*, p.328.

As for Chalmers' hypothetical, given the enormous redundancy apparent in the human brain, one would expect that the replacement or elimination of rather large numbers of neurons without diminishing consciousness, but there could still be that one neuron that "breaks the camel's back." In addition, while Chalmers' dismisses the possibility of "fading qualia" (the qualities of experiencing), I find it easy to imagine the gradual loss of mental functionality until, at some point, we would say consciousness is gone. So, the hypothetical proves nothing to me.

"This seems to be quite implausible. Here we have a being whose rational processes are functioning and who is in fact conscious, but who is utterly wrong about his own conscious experiences. Perhaps in the extreme case, when all is dark inside, it might be reasonable to suppose that a system could be so misguided in its claims and judgments—after all, in a sense there is nobody in there to be wrong. But in the intermediate case, this is much less plausible."

Id., p.257.

I have pondered a different hypothetical, one related to the Star Trek transporter we have discussed. Suppose that that technology enabled the generation of two Captain Kirks. Would they both be conscious? Would they have identical memories? Would their subsequent experiences feel the same? Some might answer these questions in the affirmative. But, if that result seems wrong, which Kirk would be conscious? The other, to me more likely, possibility is that neither would be conscious, that the transporter could send a body elsewhere, but not a mind, not consciousness.

I doubt that AI will become conscious.

First, artificial intelligence (AI) is based upon preprogrammed mechanical steps and brute computational force. The device picks a possible solution, calculates the amount of error, makes a minor adjustment to the proposed solution, calculates the error and repeats the process over and over again until the calculated error is within the preselected range. The result will be a satisfactory solution, not necessarily the best solution.

"Is identifying correlations the same thing as thinking and reasoning?" Anil Ananthaswamy, *Why Machines Learn: The Elegant Math Behind Modern AI* (2024), p.25.

Large Language Machines (LLMs) like ChatGPT calculate from a data base the probabilities that a word or phrase has been followed by a particular word and "predicts" what the next word in a sentence would be, but today the data bases are enormous, far exceeding what a human could absorb.

"Once trained, the LLM is ready for inference. Now given some sequence of, say, 100 words, it predicts the most likely 101st word. (Note that the LLM doesn't know or care about the meaning of those 100 words: To the LLM, they are just a sequence of text.) The predicted word is appended to the input, forming 101 input words, and the LLM then predicts the 102nd word. And so it goes, until the LLM outputs an end-of-text token, stopping the inference."

Ananthaswamy,, pp.418-419.

"[A]n LLM is trained on an internet's worth of data; no child will ever come remotely close to experiencing so much language during learning. Still, LLMs can learn syntax and grammar from the statistical patterns that

---

exist in human written language, and they have some notion of semantics. To an LLM, the word 'heavy' might not mean quite the same thing as it might to humans, yet the LLM is able to 'reason' about heaviness and lightness in ways that suggest at least some semantic understanding of those words. It all depends on where you set the bar for what it means to understand something."

Id., pp.428-429.

"Both PaLM and Minerva were trained using self-supervised learning, meaning they were taught to predict masked tokens in some sequence of tokens that appeared in the training data. They were not taught to explicitly reason or solve math problems. ...It takes the question, turns it into a sequence of tokens, and then simply predicts what follows, token by token. Out comes what appears to be a reasoned answer. Is Minerva simply regurgitating text based on correlations in the training data? Or is it reasoning? The debate is raging, and no clear answers are forthcoming. ...The theory isn't sophisticated enough to resolve the debate. The experiments themselves don't substantiate claims one way or another; they simply provide evidence that needs explaining."

Id., pp.412-413, 414.

The point is that AI does not do what the brain does. Observe the learning of a child from age 2 to 5. The process is obviously vastly more efficient than that utilized by AI. The problem is that we do not know how the brain does it. There must be "shortcuts," special techniques and/or extraordinary abilities, all mysteries to us. A small child "divines"or "intuits" the meaning of words, presumably from the contexts in they are being used. It is clearly something more than noting correlations or detecting patterns. Perhaps, some innate mental capabilities are involved. We don't know.

"Newborn ducklings, with the briefest of exposure to sensory stimuli, detect patterns in what they see, form abstract notions of similarity/dissimilarity, and then will recognize those abstractions in stimuli they see later and act upon them. Artificial intelligence researchers would offer an arm and a leg to know just how the ducklings pull this off." *Id.*, p.8.

Second, AI requires the programmer to select the test data and to specify the complexity level to be used. The selection of the test data can bias the device's output. The complexity level will affect the accuracy of the predictions.

"Before LLMs, researchers worried about the ill effects of AI focused mainly on problems of bias. ...ML algorithms assume that the data on which they have been trained are drawn from some underlying distribution and that the unseen data on which they make predictions are also drawn from the same distribution. ... [T]hey must explicitly de-bias the data, to ensure that the algorithm's predictions are accurate. They must also ensure they are asking the right questions of the data."

Id., pp.420, 422.

"This leads us to the two competing forces at work here. One is called bias: The simpler the model, the greater the bias. The other is called variance: The more complex the model, the greater the variance. High bias (i.e., simpler models) leads to underfitting, a higher risk of training error, and a higher risk of test error, whereas high variance (i.e., more complex models) leads to overfitting, a lower risk of training error, and a higher risk of test error. The job of an ML engineer is to find the sweet spot."

Id., p.389.

"It's ... unclear whether pure machine learning, or learn-ing about patterns in data, can really get us to the kind of intelligence that biological brains and bodies demonstrate. Our brains are embodied. Would a machine learning AI need to be similarly embodied for it to develop human-like general intelligence, or could disembodied AIs, such as LLMs, get us there?"

Id., pp.429-430.

"Are LLMs reasoning? ... Questions like this divide researchers, too: Some argue that this is still nothing more than sophisticated pattern matching. ... Others see glimmers of an ability to reason and even model the outside world. Who is right? We don't know, and theorists are straining to make mathematical sense of all this." *Id.,* p.420.

In the end, Chalmers relies on our lack of knowledge to support his conclusion:

- "[T]here seems to be no good argument for the thesis that computational dynamics at a basic causal level is incompatible with creativity and flexibility at the macroscopic level." Chalmers, p.329.
- "Certainly it is not obvious that executing the right sort of computation should give rise to consciousness; but it is not obvi-ous that neural processes in a brain should give rise to conscious-ness, either." Chalmers*Id.,* p.314.

I disagee with the first statement. It could read "compatible with" and be more consistent with our experience. Imagination and creativity are not ubiquitous. They are actually quite rare and unusual. We would not expect to find them, and we would generally doubt their existence absent evidence of their presence.

The second statement is correct, except that "neural processes in a brain" do, in fact, do so We know what does work, but not what might work. Only guesses.

"If we take our image of the population, speed it up by a factor of a million or so, and shrink it into an area the size of a head, we are left with something that looks a lot like a brain, except that it has homunculi—tiny people—where a brain would have neurons. On the face of it, there is not much reason to suppose that neurons should do any better a job than homunculi in supporting experience. **Of course, as Block points out, we know that neurons can do the job, whereas we do not know about homunculi.**"

Id., p.252 (emphasis added)

We have not disproven dualism yet. Thus, it may be that the functional invariance principle simply does not apply to consciousness.

"On the view I advocate, consciousness is governed by natural law, and there may eventually be a reasonable scientific theory of it. There is no a priori principle that says that all natural laws will be physical laws; to deny materialism is not to deny naturalism. A naturalistic dualism expands our view of the world, but it does not invoke the forces of darkness."

Id., p.170.

"Nothing about the dualist view I advocate requires us to take the physical sciences at anything other than their word. The causal closure of the physical is preserved; physics, chemistry, neuroscience, and cognitive science can proceed as usual In their own domains, the physical sciences are entirely successful. They explain physical phenomena admirably; they simply fail to explain conscious experience."

Id., p.170.

"We know about our own detailed and specific conscious experiences, and we also know about the underlying physical processes, so there is a significant set of regularities right there. Given these regularities, we can invoke some sort of inference to the best explanation to find the simplest possible underlying laws that might generate the regularities."

Id., p.216.

"[W]hile this theory will not explain the existence of consciousness in the sense of telling us 'why consciousness exists,' it will be able to explain specific instances of consciousness, in terms of the underlying physical structure and the psychophysical laws. Again, this is analogous to explanation in physics... ."

Id., p.214.

CHAPTER 6

Causation

"You ask what is the use of classification,
arrangement, systematization? I answer you:
order and simplification are the first steps
toward the mastery of a subject—
the actual enemy is the unknown."

...

"Nothing in the domain of life seemed
uncausated, or insufficiently causated,
from that point on; but life itself seemed
without antecedent."

Thomas Mann
The Magic Mountain
(1924)
(Kindle Edition, Loc.4717, 5292).

"As Hume argued,
external evidence only gives us access
to regularities of succession between events;
it does not give us access to any further fact of
causation. So if causation is construed
as something over and above the presence
of a regularity (as I will assume it must be),
it is not clear that we can know that it exists.

...

"We infer the existence of causation by a kind
of inference to the best explanation—
to believe otherwise would be to believe
in vast, inexplicable coincidences—... ."

David J. Chalmers
The Conscious Mind:
In Search of a Fundamental Theory
(1996), pp.74-75.

I.

I have discussed many issues arising from the concept of causation. *Important Things We Don't Know About Nearly Everything*, pp.36-64; *Imaginings,* pp.55-102. The idea of causation is a fundamental element of the human mind. Obviously, early humans noticed and took note of apparent regularities in nature. Doing so reduced uncertainty. But, many of our uses of the concept of causation embody notions of intent.

One of the more basic question is whether causation exists in the microscopic world. Of course, we believe that certain predictable outcomes necessarily occur following particular interactions, but what constitutes causation? From the microscopic elements constituting our

Universe, science tells us how that Universe emerged and behaves. And, although some scientists shy away from the word "why," science provides answers to many of the "why" questions. For example, why does the apple fall from the tree? Because of gravity. That does pretty well as a response to why, but scientific theories are less compatible with our conception of causation.

Is gravity the "cause" of the apple's fall? Something occurred so that the stem became detached. In addition, the mass of the apple is a necessary element, as is the mass of the Earth. Throw a ball into the air. What is the cause of it falling? Gravity? Or, do we look at how it got into the air to return to the ground where it began. We could describe the event in terms of potential energy, inertia, an external force, kinetic energy, air resistance and gravity. So, the "cause" is... ?

Nevertheless,

> "Many scientists assume that all causation occurs at the most microphysical layers of reality, at the fundamental scale of elementary particles. This leaves no room at large scales for what is called higher-level causation... ."

Sara Imari Walker, *Life as No One Knows It: The Physics of Life's Emergence* (2024), p.23.

Now, that sure seems like the tail wagging the dog.

Something is not right.

I may be missing some subtleties of the argument, but I find it inconceivable to conclude that consciousness has no effects on the world. Consciousness enables the imagination of things that do not but could exist, of events that have not but could occur and of solutions to unsolvable problems. These mental phenomena—in fact—do lead to intentions, which—in fact—do lead to actions that alter, often quite dramatically, our world.

"So we need to ask, does anything in the universe exist that might not be possible if subjective inner worlds did not exist? The answer, I think, is yes. There are things like rockets, which existed within our collective imagination for centuries before they were ever built as physical objects with definite properties."

Walker, p.46.

"[T]he key feature is the ability to imagine or represent things that do not exist, but could, such that the act of imagination becomes causal to the existence of some objects." Walker, p.73.

Here is an illustration that I used before borrowed from Terrence Deacon:

"A boy picks up a flat, smooth stone and throws it so that it 'skips' across the surface of the lake. *Incomplete Nature*, pp.18–22. The boy presumably was taught how to accomplish this feat by someone else. Because of that instruction or example, he was able to visualize the possibility of the event and take steps to realize it. The result was a stone 'skipping' on water. '[P]rior to the evolution of humans, the probability that any stone on any beach on Earth might exhibit this behavior was astronomically minute.' *Id.*, p.20. Something happened that changed the way in which matter behaved and interacted.

"What was the 'cause' of the stone's unusual behavior? Various people could describe in painful detail the bases in classical mechanics for the fact that the stone 'skipped,' given its shape, size, speed and trajectory [angle of impact]. But, what 'caused' the phenomenon? Certainly, the boy's decision to try to skip the stone was a 'cause,' and one that 'explains' the shape, size, speed and trajectory of the stone. And, that decision appears to have been the result of his ability to visualize or imagine a future result and his understanding of how to bring about that result. *See id.*, pp.21–2."

Important Things, pp.587-8.

The obvious additions to the basic physics are imagination, intention and agency.

Indeed, a "cause" suggests a "why." Today, we are likely to object to "why" questions as unscientific. Our investigations are into "how" something comes to be and "what" the something is, but avoid asking the "why."

"'Why?' is not really a sensible question in science because it usually implies purpose... ." Krauss, (Kindle Edition) Loc.90.

"[I]n science we have to be particularly cautious about 'why' questions. When we ask, 'Why?' we usually mean 'How?' If we can answer the latter, that generally suffices for our purposes." Krauss, p.143.

The existence of causal impacts is not the real issue. The troublesome questions are (i) how matter causes imagination, discovery, inspiration, insight, and intention and (ii) how such mental states are then translated into physical acts. It is not whether they do, but how they do that is the puzzle. Determinism avoids this lack of understanding the how by denying that it happens. I do not find that a credible position.

II.

> "The physical world is more or less causally closed,
> in that for any given physical event,
> it seems that there is a physical explanation... .
> This implies that there is no room for a nonphysical
> consciousness to do any independent causal work.
>
> ...
>
> "It exists, but as far as the physical world is concerned
> it might as well not."
>
> David J. Chalmers
> *The Conscious Mind:*
> *In Search of a Fundamental Theory*
> (1996), p.150.

The zombie hypothetical raises a host of issues and also provides a platform for exploring others. Here, I consider the matter of consciousness and causation. At one end, one can conclude that consciousness has no effect on the physical world, causes nothing to happen. As I define consciousness, that position necessarily implies the absence of free will. Every action is determined by prior events and external factors.

Alternatively, if conscious persons make decisions sometimes based on (causeed by) feelings, then what about the zombies? If the behavior of the zombie twin is identical to that of the conscious twin, as specified in the hypothetical; then, there must be something other than consciousness causing the zombie to make the same decisions as its twin. It could be that there are two causes for the concious twin's decision, or we can conclude that the zombie hypothetical is not conceivable. I am inclined toward the latter conclusion, but that conclusion undermines one of the arguments that consciousness is something more than the physical. Certainly, it does not establish the opposite. But, we are faced squarely with the question of how the immaterial can affect the physical world.

(Of course, "the very nature of causation itself is quite mysterious, and it is possible that when causation is better understood we will be in a position to understand a subtle way in which conscious experience may be causally relevant." Chalmers, p.150.)

Chalmers suggests that "the phenomenology of experience in human agents may inherit causal relevance from the causal role of the intrinsic properties of the physical." *Id.*, p.154.

> "Intuitively, it is ... reasonable to suppose that the basic entities that all this causation relates have some internal nature of their own, some intrinsic properties, so that the world has some substance to it. But physics can at best fix reference to those properties by virtue of their extrinsic relations; it tells us nothing directly about what those properties might be."

Id., p.153. *See below*, "Reality."

Perhaps, he suggests:

> "The world ... consists in a vast causal network of phenomenal properties underlying the physical laws that science postulates. ...The basic properties of the world are neither physical nor phenomenal, but the physical and the phenomenal are constructed out of them. From their intrinsic natures in combination, the phenomenal is constructed; and from their extrinsic relations, the physical is constructed."

Id., p.155.

Such a theory, if it comes to exist, would accomodate causality.

CHAPTER 7

Agency

"I hope ... you have nothing against malice?
In my eyes, it is reason's keenest dart
against the powers of darkness and ugliness.
Malice, my dear sir, is the animating spirit
of criticism; and criticism is the beginning
of progress and enlightenment."

...

"What a creature is man,
how widely his conscience betrays him!
How easy it is for him to think he hears,
even in the voice of duty, a licence to passion!"

Thomas Mann
The Magic Mountain
(1924)
(Kindle Edition, Loc.1146, 3033).

I have a particular interest in what I will call agency, the belief that one is choosing, is intentionally pursuing a goal, is master of one's fate (at least, to some extent).

There are two separate phenomena involved here: first, what creates the perception of choice that exists, even if one assumes that there is no free will and the outcome was predetermined; second, how is a decision made, if one assumes that free will does exist. (I am referring to the freedom to make irrational or arbitrary decisions, something that cannot be preprogrammed or predicted. That is what conflicts with determinism.) We have no answer for either.

> "Agency roughly corresponds to cases where we think **a system appears to have its own causal control over what happens and is not subject to the whims of the environment** around it. **Goals allow agents to steer and choose among possible futures.** Humans can be said to exhibit goal-directed behavior **because we imagine possible futures** and select among them in terms of how we act... ."

Walker, p.217 (emphasis assed).

> "As technical terms [agency and free will] have defied ready definitions or explanations because right now they are based more on our experience of the world than what is captured by our scientific understanding of it. ...[Some science] removes free will as a possibility at all, because it assumes causation (without control) is sufficient to describe everything. This view ignores that agents also can influence what happens.
>
> ...

"[But] by invoking agency as a causal category, there is still an open question: we do not know what gives agents this special causal power of control."

Walker, pp. 21, 24-25.

A belief in dualism is understandable—the human perception is of the mind as something separate and distinct. Yes, we do identify with our bodies and often feel that our bodies are us. But, especially when our bodies malfunction, we became acutely aware that our mind functions in some important way independently from the body. Sharp pain, a sudden substantial drop in blood pressure or severe nauseous can overwhelm the mind and push it out of the way, and the brain (and, presumably, the mind) clearly depends on the body for survival. But, similarly, a tree is dependent on water, the sun and the soil in which it grows, yet it has an existence of its own. To the experienced driver, the car can feel like an extension of one's body. Until it breaks down.

Of course, we now understand that the brain is the physical location of the mind and, in many ways, does its own thing independent of the body. The philosophical question concerns the relationship between the brain and the mind. Do the physical interconnections of the brain "cause" what happens in the mind? Or, is there something more involved?

In particular, how are decisions made, assuming that they are not all hardwired. And, how do they manage to affect physical events.

"[I]f consciousness is a material property of at least some objects, what does it do or cause in the world that we might observe only for those objects that have it? Does this allow us to see something deeper that explains what consciousness is and why it exists? ...By focusing on whether consciousness causes anything we could not explain otherwise, we re-center the problem of consciousness on something measurable. ...If consciousness is a real thing, and not just an illusion or epi-phenomenon, as some people will claim, then it should have some measurable impact on how a physical system behaves."

Walker, pp.44-45, 46.

II.

**"Knowledge ... allows physical systems
to be able to experience the world**
and to generate patterns of activity that have
meaning and feeling,
thereby transforming nonliving parts of the universe
into **pieces with purpose."**

Seyed B. Azarian
The Romance of Reality:
How the Universe Organizes Itself to Create
Life, Consciousness, and Cosmic Complexity
(2024), p.155 (emphasis added).

Azarian's "theory of everything" is based upon the concept of information. He uses it to address the question:

"How exactly did molecules acquire agency, control, and purpose? Precisely what process instilled a physical system with goals and the capacity to pursue them?"

Azarian, p.58.

The source or cause of agency, control, and purpose?

That is the question. (And, of course, do cells and micro-organisms have discretion? Do they really choose among options? Make decisions?)

"Could the origin of agency be the key to understanding the emergence of free will and consciousness—two of the great scientific mysteries of our time?"

...

"The origin of life, wherever it occurred first, was the moment when **information gained causal power** in the universe. Inanimate systems may be able to be described in terms of information, **but information becomes meaningful to reality when it can make a difference to an agent.**"

Id., pp.79, 213 (emphasis added).

Information—the key to agency?

A focus on information has previously been used as an argument supporting design. Azarian presents a twist on that argument, however, by accepting the implication of design, but finding design in the supposed purposefulness or direction in the cosmos. *See infra.,* pp.148-53.

"Complex objects do not come into existence without the information first existing to make them. The question then is where this information comes from—is it designed or evolved, or both *(e.g.,* by evolution generating things that are capable of design—things like us)? ...While physicists think that their theories refute intelligent design, the idea of the spontaneous formation of any object anywhere in the universe ... is the greatest argument for it: it implies every point in space and time contains the design of every object."

Walker, pp.70, 113-114.

"The relatively recent science of information has given rise to the new formulation of many of the arguments we have been discussing, on issues from biology to physics. See Stephen C. Meyer, Signature in the Cell: DNA and the Evidence for Intelligent Design (2009); Charles Seife, Decoding the Universe, p.385. ...The key to life, as we understand it, is the reproducible genetic code contained in RNA and DNA. So, the need for a designer in this context arises not from the complex or sophisticated ('designed') nature of organs or organisms, but from the appearance of the coded genetic information that enables life. It is 'the mystery of the origin of the information needed to build the first living organism.' Meyer, Signature in the Cell, p.14."

...

"Meyer asserts that: 'Intelligence is the only known cause of complex functionally integrated information-processing systems.' Id., p.346."

Important Things We Don't Know, pp.525, 528.

Azarian's argument relies on natural selection, focusing on a model of self-correcting, "trial and error" processes—a version of natural selection that goes to work on knowledge or, more precisely, on the physical structures that embody knowledge leading to wonderous results.

"[A] self-correcting system grows more robust and computationally power-ful because it is always solving survival problems and storing the solutions in memory." *Id.*, p.115

"As the brain encodes adaptive information ..., our mental model of the world, the representation of reality that defines the mind, gets built up and updated continuously... ." *Id.*, p.90.

"This interpretation of the evolutionary process stresses that adaptive systems store a simplified representation or model of the world they inhabit. As evolution and adaptive learning proceed, life is effectively performing statistical inference and updating its model's 'beliefs about the world" *Id.*, p.118

"Essentially, mutation is an inventor and natural selection is a pruner. The genetic and neural information that is left over after natural selection culls the generated biodiversity reflects knowledge of the environment that enables the ordered system to survive and thrive." *Id.*, pp.130-131.

"As external conditions change over time, new adaptive solutions are discovered and written into genetic memory, such that new knowledge is continuously being generated, along with new modes of behavior." *Id.*, p.131.

The heart of his argument is as follows:

1. "[I]t is the capacity to acquire, store, process, and transmit information that separates life from nonlife." *Id.*, p.62.

2. "If uncertainty is mathematically equivalent to disorder, and information is a reduction in uncertainty, then information acquisition must be associated with a reduction of disorder." *Id.*, p.66. And, "[k]nowledge is the information we acquire that reduces our uncertainty or ignorance about the world." *Id.*, p. 84.

3. Knowledge or information is embedded in or captured by physical structures, including "...DNA, life's original memory system. ...[G]enetic memory accumulates in an evolving population of organisms over many generations, as a result of replication with mutation and natural selection... ." *Id.*, p.92.

4. These structures are subject to natural selection. (The physical structures. Sometimes, Azarian seems to be arguing that knowledge or information in subject to natural sselection and evulution, which is not correct.)

Thus:

- "As the brain encodes adaptive information in the synaptic circuitry created by networks of neurons, our mental model of the world, the representation of reality that defines the mind, gets built up and updated continuously through a process roughly analogous to Bayesian updating."
- "According to the alleged 'unified brain theory,' sentient systems naturally try to resolve environmental uncertainty by behaving in ways that minimize prediction error—or the amount of 'surprise' an agent experiences in its encounters with the world—which makes the mental model more accurate."
- "Knowledge creation through blind variation and natural selection elegantly solves the problem of epistemology and the problem of teleology by providing an intelligible mechanism for how organisms come to possess accurate internal representations (knowledge) of the external world... ."
- "An agent acts on its environment, observes the change in that environment, updates its model by encoding this change in memory, and through iterations of this process, the organism's world model creates a data variable for itself."

- **"Agency only requires that a system have a world model**—an internally embedded statistical mapping that represents the biologically relevant regularities in the environment."

Id., pp.90, 91, 95, 220, 217-218 (emphasis added).

Sara Imari Walker also stresses the importance of memory and natural selection to agency. Memory enables the contemplation of alternatives and the formation of goals. Walker says: "A living cell is 'just' a bag of chemicals that is animated by the memory it stores and the goals it acts on." *Life as No One Knows It: The Physics of Life's Emergence* (2024), p.223.

"If an object includes a history that allows it to construct a future it persists in, then it can pass the test. If it cannot do this, the object will display no behaviors that cannot already be accounted for in current physics. Here the memory is the object." Walker, p.220.

"To model the future, you must have a past. ... You cannot have goals unless you have a system that was selected. In fact, if assembly theory is on the right track the mechanisms of selection and goal-directed behavior are one and the same: the only difference is whether we are looking at the mechanism backward in time (selection) or forward (goals). To make goal-directed matter you therefore need selection."

Walker, pp.221-222.

But, she also does not attempt to explain how the alternatives are evaluated or choices are made. Moreover, there is no explanation of the cause or consequences of imagination or creativity or of new ideas (inspirations). That gap is particularly significant given that much of the creative work appears to occur in the subconscious mind. *See Imaginings*, pp.155-70.

So?

First, it is troubling to describe the evolution of a bacteria as the acquisition of knowledge. It is even more problematic to characterize the adaptive mutations as being ever-improving "predictors."

"As a species adapts to a niche over many generations, the design and behavior of the prototypical organism becomes more statistically correlated with its environment, and therefore it **becomes a better predictor** of that environment." *Id.*, p.132 (emphasis added).

Second, while the analogy of an increasingly accurate mental model of the environment makes some sense with respect to conscious entities, it is seriously misleading with respect to unconscious life forms. They simply do not compare and select. It happens to them, not by them.

Third, a crucial assumption is that physical changes through trial and error alter the state of the embedded knowledge and that improvements in the knowledge increase the survivability of the host organism. Perhaps. But, I doubt that there is a one-to-one correspondence between physical changes and increased knowledge.

Fourth, what could be the source and content of the original model of the environment that is then being improved?

Fifth, memory must be available for natural selection to operate. So, from where does it come?

Finally, Azarian's theory does not explain the origin of RNA or DNA or of life.

Oh, and, we still have no explanation as to how supposed decisions are made.

"...Hofstadter believes that the self, the I, the conscious observer, has true causal power on the world. He also gives us a vague but powerful insight into how this observer capable of causation comes into existence. By his account, the self emerges from self-reference 'via a kind of vortex whereby patterns in a brain mirror the brain's mirroring of the world, and eventually mirror themselves, whereupon **the vortex of 'I' becomes a real, causal entity.'"**

Azarian, p.229 (emphasis added).

"We can actually **redefine life as a system with the ability to initiate new causal chains**. Causal chains are chains of cause and effect that result from an agent's actions on its environment." *Id.*, p. 213 (emphasis added).

"What makes the idea in this book different, and not just different, but superior... ?" *Id.*, 82.

Good question.

As an aside, Azarian asserts that this theory solves the supposed problem of the tautological nature of Darwin's natural selection based on survival of the fitest, in which fitness is defined by the fact of survival. I have previously discussed my view that that apparent tautology is not really problematic (*Important Things We Don't Know*, pp.222-3), but let's look at the proposed solution:

"The integrated evolutionary synthesis gets rid of the fitness tautology... . From the thermodynamic perspective, fitness pertains to how well an organism can resist entropic decay. In particular, fitness corresponds to the

resilience of the embodied energy-extraction program, the dissipative chan-
nel, and is not correlated with strength, intelligence, or even complexity."

Id., p.137.

So, we can say "survival of the most resilient embodied energy-extraction
program." Not very catchy. And, hardly less circular. "Resilience" is still not
defined independent of the outcome of survival. It does not have its own
measurement scale like, for example, "the tallest" or "the strongest."

III.

These theories are seriously inadequate. Memory and information
may well be necessary for agency, but they are hardly sufficient. Suppose
an organism can remember alternative scenarios from its past. How
does it choose? What is the process that occurs? Perhaps, it also has
information about its current environment and "selects" the scenario
that best matches the environment. Has it made a "choice" or simply
matched a pattern, following mechanical rules implanted through natural selection? More importantly, agency in conscious entities derives
from imagination, inspiration and feelings. Where do those factors
come from and how are they utilized?

More questions than answers. Many more.

CHAPTER 8

Emergence

"Thus in the inmost recesses of nature,
as in an endless succession of mirrors,
was reflected the macrocosm of the heavens,
whose clusters, throngs, groups, and figures,
paled by the brilliant moon, hung over
the dazzling, frost-bound valley... ."

Thomas Mann
The Magic Mountain
(1924)
(Kindle Edition, Loc. 5468)

"The remarkable progress of science ... has given us
good reason to believe that **there is
very little that is utterly mysterious about the world**.
For almost every natural phenomenon
above the level of microscopic physics,
there seems in principle to exist
a reductive explanation: that is, an
explanation wholly in terms of simpler entities.
In these cases, **when we give an appropriate account
of lower-level processes, an explanation
of the higher-level phenomenon falls out**."

David J. Chalmers
The Conscious Mind:
In Search of a Fundamental Theory
(1996), p.42 (emphasis added).

"David Chalmers ... consider[s] that the only thing that comes
close to qualifying as a case of strong emergence is consciousness; ...
Sean Carroll ... thinks that while consciousness is the only real reason
to be interested in strong emergence, it's sure not a case of it."

Robert M. Sapolsky,
Determined:
A Science of Life without Free Will
(2023), p.197

The word "emergence" is used to "explain" a lot. But, such claims need to be examined carefully and individually. The attachment of that label does not explain how or why something happens. Indeed, emergence seems to be a sscientist's way of saying something happened that I cannot explain.

For example:

"Much of life is emergent. Its organization and behavior cannot be predicted from the way its parts come together; rather it is a consequence of their interactions. ...Behind all of these biological organizations and structures is the cell, which itself should be seen as an emergent structure."

...

"From these circuits emerge our mind and consciousness. While genes provide instructions for the proteins needed to build these circuits, it is the cells that interpret ensembles of proteins to hook up with each other in these different ways. The circuits are the products of interactions among cells, not genes, and their output is another, well-accepted example of emergence."

Alfonso Martinez Arias, *The Master Builder: How the New Science of the Cell Is Rewriting the Story of Life* (2023), pp.94, 95, 130.

"[E]mergence ... us[es] simple rules that work in similar ways in solving optimization challenges for, among other things, ants, slime molds, neurons, humans, and societies."

...

"[E]mergence helps slime molds solve problems."

...

"[C]omponents of a system interact locally, repeated a huge number of times with huge numbers of those components, and out emerges optimized complexity. All without centralized authorities comparing the options and making freely chosen decisions."

Sapolsky, *Determined,* pp.159, 163, 178-179.

"If, as the universe cooled, it underwent some kind of phase transition—as occurs, for example, when water freezes to ice or a bar of iron becomes magnetized as it cools—then not only could the Horizon Problem be solved, but also the Flatness Problem... ."

Lawrence Krauss, *A Universe from Nothing: Why There Is Something Rather than Nothing* (2012), p.95.

"Many have wanted to reject a reductive account of consciousness while giving it a central causal role. A popular way to do this has been to argue for emergent causation—the existence of new sorts of causation in physical systems of a certain complexity."

David J. Chalmers, *The Conscious Mind: In Search of a Fundamental Theory* (1996), p.378.

Sapolsky characterizes the various collective phenomena as examples of "emergence." Robert M. Sapolsky, *Determined: A Science of Life without Free Will* (2023). Most of these collective phenomena, of course, are irrelevant to or make no sense with respect to individual cells or organisms. A single molecule cannot feel wet or hot or cold. A single cell cannot "ooze" or communicate. A single cell cannot follow another. It cannot be a solid or a liquid. These are all things that require more than one, even a collective. But, that does not necessarily make them examples of emergence. *See infra.*, pp.131-42.

If emergence itself is considered to be something immaterial, then it does appear to provide an answer to one fundamental question—how does the immaterial cause behavior of material things? The properties of the emergent consciouness are material things.

This concept can include "downward causation," where the higher level state resulting from the combination of lower level building blocks causes changes in those building blocks. (This concept is discussed elsewhere. *See Imaginings,* pp.58-6, 63-71.) "[W]e all agree that altering the little can change the emergent big. And the reverse certainly holds true... . ...[S]ome emergent states have downward causality, which is to say that they can alter reductive function and convince a neuron that long is short... ." Sapolsky, p.198.

But, compare:

"[T]he whole point of emergence... is that those idiotically simple little building blocks that only know a few rules about interacting with their immediate neighbors remain precisely as idiotically simple... . Downward causation doesn't cause individual building blocks to acquire complicated skills; instead, it determines the contexts in which the blocks are doing their idiotically simple things." Sapolsky, pp.200-01, 202.

"The emergence of top-down causation is an example of what philosophers call strong emergence, because there is a fundamental change in the kind of causality we see in the world. ...Top-down causation is ... a mathematically formalized mechanism that describes how a system with control (like an organism or a brain) can steer itself toward goals, thereby influencing the paths of its low-level components (the molecules ...)." Bobby Azarian, *The Romance of Reality: How the Universe Organizes Itself to Create Life, Consciousness, and Cosmic Complexity* (2024), pp.212, 213.

"[T]here is no evidence for such emergent principles of causation. As far as we can tell, all causation is a consequence of low-level physical causation, and 'downward causation' never interferes with low-level affairs. Second and perhaps more important: on a close analysis, the view leaves consciousness as superfluous as before." Chalmers, pp.378-379).

So, what are emergent phenomena?

The visibility of a mass of cells individually too small to see is not an emergent phenomenon. Emergence is where something appears or occurs in the presence of the collective that is not observable in nor predictable from the individual entities.

Martinez Arias explains (p.93):

> "The first basic element of an emergent structure or process is that **the components of the system must interact.** ...[W]hen these elements come together and combine with each other, **they exchange information** about their states or structure. ... The second element of an emergent process is that the interactions follow very simple rules... . Importantly, **the outcome cannot be predicted** merely from knowledge of the components... ." (Emphasis added.)

"To understand how the parts of a cell come together to generate structures that can do more than one can glean from their individual makeup, we need to understand one important concept: **emergence**, a phenomenon whereby **a combination of parts leads to activities not found in any of the individual parts**. ... The phenomenon whereby interactions between the component parts of a whole ... produces structures and behaviors that **cannot be predicted from the assembly of its individual parts** is called **emergence**."

Martinez Arias, p.92 (emphasis added).

Sapolsky agrees: "[E]mergence is about reductive piles of bricks producing spectacular emergent states, ones that can be **thoroughly**

unpredictable or that can be predicted based on properties that exist only at the emergent level." Sapolsky, p.192 (emphasis added).

"What makes for emergent complexity?

"—There is a huge number of ant-like elements, all identical or coming in just a few different types. ...

"—There are a few simple rules based on chance interactions with immediate neighbors (e.g., 'walk with this pebble in your little ant mandibles until you bump into another ant holding a pebble, in which case, drop yours'). No ant knows more than these few rules, and each acts as an autonomous agent.

"—Out of the hugely complicated phenomena this can produce emerge irreducible properties that exist only on the collective level."

Sapolsky, pp.156-157.

Perhaps, we can think of "weak" emergence as a phenomenon of the human mind's efforts to model and understand the world. *See Imaginings*, pp.60-73. We construct theories of collective or macro-level behavior that do not diirectly derive from nor are predicted by the characterizations of individual or micro-level events, but that are consistent with and explainable by them after †he fact.

"Ontological [strong] emergence, in the sense that matters here, concerns the emergence of genuinely novel properties at some non-fundamental level, while epistemic [weak] emergence concerns the emergence of greater explanatory or predictive power at some higher level of description, relative to our epistemic capacities." Joe Dewhurst, "Causal Emergence and Real Patterns," *philsci-archive.pitt.edu*, April 3, 2020.

"If microphysical facts **entail** a high-level fact, this is because the microphysical facts **suffice to fix those features** of the world in virtue of which the high-level intension applies. That is, we should be able to analyze what it takes for an entity to satisfy the intension of a high-level concept, at least to a sufficient extent that we can see why those conditions for satisfaction could be satisfied... ."

...

"[T]he relevant properties are not obvious consequences of low-level laws; but they are still logically supervenient on low-level facts. If all the physical facts about such a system over time are given, then the fact that self-organization is occurring will be straightforwardly derivable. ...[J]ust what we would expect, as properties such as self-organization and flocking are straightforwardly functional and structural."

Chalmers, pp.77, 129 (emphasis added).

"Strong" emergence, in contrast, is something that occurs in the collective without any identifiable connection to the individual elements, even with hindsight. Something truly new.

> "If consciousness is an emergent property, it is emergent in a much stronger sense. There is a stronger notion of emergence, ..., according to which emergent properties are not even predictable from the entire ensemble of low-level physical facts. ...But this sort of emergence is best counted as a variety of property dualism. ...[T]he strong variety requires new fundamental laws in order that the emergent properties emerge."

Chalmers, pp.129-130.

"These views should not be confused with the 'innocent' view of emergent causation found in complex systems theory, on which low-level laws yield qualitatively novel behavior through interaction effects. On the more radical view, there are new fundamental principles at play that are not consequences of low-level laws." Chalmers, p.378.

So, are the phase transitions from gas to liquid to solid "weak" or "strong" emergence?

The random motion of a molecule changes with temperature. As the temperature rises, the motions of the molecule increase. Thus, it is not surprising that the solid structure of ice would breakdown as the temperature rises, with the more rapid movements of the water molecules resulting in a liquid. Higher temperatures will result in the liquid becoming a gas. So, the observed phases of the collective of molecules can be explained by characteristics of the individual molecules.

But, why do the transitions occur at the specific temperatures rather than others ? Why do they occur suddenly rather than gradually? Why do they occur in reverse as temperature declines? These are characteristics of the collective, apparently neither explainable not predictable from an examination of an individual water molecule.

Superconductivity?.

Various aspects of the functioning of cells do appear to arise when the cell is in a particular location or position relative to other cells—phenomena of the collective. Perhaps, further investigations will reveal mechanisms in the individual cells that can be used to explain the collective behavior. We might thereby "discover away" strong emergence for many other phenomena. Or not.

The challenges arise in chemistry, where there is still a significant gap between our theories of atoms and the physical characteristics of molecules as matter: *e.g.*, color, smell, taste, hardness, phases and the phase transitions, and in biology, including the behavior of cells and neuroscience.

"[T]he complexity of chemistry makes the reduction to tests from fundamental principles far more difficult. Might progress in establishing the basis for the concepts of chemistry lead to consistency with present standard physics or, even better, to an improved fundamental theory? It is a fascinating prospect to be checked as methods of ab initio chemistry improve."

P. J. E. Peebles, *The Whole Truth: A Cosmologist's Reflections on the Search for Objective Reality* (2022), pp.204-205.

Presumably, we will continue to use weak emergence to refer to characteristics that only appear in the collective. Many of these characteristics will be structural.

Some other phenomena referred to as emergent, probably are not.

For example, Sapolsky's descriptions of apparent "problem solving" attribute the behavior as resulting mechanically from the "programming" of the individuals to follow a few simple rules or instructions when part of a group. Similarly, Martinez Arias describes how a particular bacteria happens to "choose" what to eat. It is a quite astonishing feat, but it occurs, as described, through mechanical chemical reactions that are effectively "preprogrammed."

"By observing 'domesticated' *E. coli* in controlled conditions in a lab, scientists have been able to discern the bacterium's economical approach: when presented with a buffet of food choices, it dines on the options in ascending order of the amount of effort eating them takes.

...

"Buried in the genome of *E. coli* are the genes—the instructions—for making the RNAs that are translated into lactose permease and ß-galactosidase, the proteins needed to process lactose. These stretches of DNA are not accessible to the cell's RNA-making machinery unless there's lactose inside the cell.

...

"Once inside the cell, the lactose binds to the repressor protein. This breaks the repressor's bond with those genes that code for the proteins to digest lactose and allows their expression. ...If there's glucose around, the cell emits a signal, in the form of proteins, that prevents the expression of ß-galactosidase ... to ensure that glucose is used first. After the lactose is consumed, the repressor is back on the job... ."

...

"Later, other circuits in *E. coli* were discovered, ...the function of each implying that it has a sensor hooked up to some sort of device for controlling gene expression."

Martinez Arias, pp. 157, 158, 159, 160.

How the programmed arrangement of proteins arose is a good question, but I would not call it an example of emergence, but simply of natural selection at the cellular level. Indeed, Martinez Arias even observes: "Genetic circuits, such as the one involved in lactose utilization, are like automated robots used to help pick inventory swiftly and efficiently off the shelves... ." Martinez Arias, p.162.

(By the way, I do not think that these are examples of "decision making" or "cooperation" either.)

What about consciousness? Can we construct a theory based upon, for example, energy conservation or states of equilibrium that connects the material of the brain with conscious experiences?

That still seems quite unlikely.

What about life?

" ...[T]here is no evidence that life violates any of the known laws of physics. But being consistent with known laws of physics does not mean life is explained by those laws." Sara Imari Walker, *Life as No One Knows It: The Physics of Life's Emergence* (2024), p.21. Walker claims that "assembly theory ... defines as fundamental all things that traditional physics would say are emergent." *Id.*, p.92.

"In the 1970s, the physicist Philip ... W. Anderson, who was awarded a Nobel Prize for his fundamental contributions to our understanding of matter, ... claimed that while we can reduce the components of the universe to elementary building blocks, and in turn describe these by reasonably simple laws (the edifice of the last four hundred years of physics), this does not imply that we can reconstruct the universe from those laws alone. Each new scale—moving up from elementary particles to atoms to chemistry to biology to technology—might have its own fundamental rules or laws, not fully re-ducible to the lower levels."

Walker, p.25.

Perhaps there will be scientific theories developed that underlay phenomena we currently label as emergence.

CHAPTER 9

Reality

"While the moon took its appointed way
above the crystalline splendours
of the mountain valley,
he read of organized matter, of the properties
of protoplasm, that sensitive substance
maintaining itself in extraordinary fluctuation
between building up and breaking down;
of form developing out of rudimentary,
but always present, primordia;
read with compelling interest
of life, and its sacred, impure mysteries."

Thomas Mann
The Magic Mountain
(1924)
(Kindle Edition, Loc.5279).

> "[Q]uestions about the world, like
> What is reality?,
> are just as central to philosophy.
> Perhaps most central of all are questions about
> the relation between mind and world, such as
> How can we know about reality?"
>
> David J. Chalmers
> *Reality+:*
> *Virtual Worlds and the*
> *Problems of Philosophy*
> (2022) p.xvi.

I have discussed previously numerous questions about the nature of reality, our ability or inability to observe it and various issues in physics like so called "action at a distance." *See Important Things We Don't Know About Nearly Everything.* However, I may not have sufficiently stressed the limitations of our best scientific theories. Yes, we do not have a theory of quantum gravity; but (as I have argued), we do not actually even have a full theory of gravity. Or, of the other forces, or of fields or, more fundamentally, even of causation.

Physics

Following an argument by philosopher David Chalmers, I note that physics is essentially about the relationships among things and how they interact with one another.

"Basic particles, for instance, are largely characterized in terms of their propensity to interact with other particles. Their mass and charge is specified, to be sure, but all that a specification of mass ultimately comes to is a propensity to be accelerated in certain ways by forces, and so on. Each entity is characterized by its relation to other entities, and these entities are characterized by their relations to other entities, and so on forever (except, perhaps, for some entities that are characterized by their relation to an observer). ...Reference to the proton is fixed as the thing that causes interactions of a certain kind, that combines in certain ways with other entities, and so on; but what is the thing that is doing the causing and combining?"

David J. Chalmers, *The Conscious Mind: In Search of a Fundamental Theory* (1996), p.153.

"We do not know what it is for any object to just be. ...When we discussed the hard problem of matter, we recognized that it is impossible to know what something is intrinsically, in the absence of interaction."

...

"Quarks are what other important particles—like the proton and neutron in atomic nuclei—are made of. But by the very physics that governs quarks (as we currently understand them), they can never exist as isolated objects and must always exist as a grouped property of the objects they make up... ."

Sara Imari Walker, *Life as No One Knows It: The Physics of Life's Emergence* (2024), pp.41, 72, 101.

Moreover,

"Things simply happen in accordance with the law; beyond a certain point, there is no asking 'how.'

...

"If there are indeed such connections, they are entirely mysterious in both the physical and psychophysical cases.... It is notable that Newton's opponents made a similar objection to his theory of gravitation: How does one body exert a force on another far away? But the force of the question dissolved over time. We have learned to live with taking certain things as fundamental."

Id., pp.170-171.

We have fashioned constructs to represent phenomena for which we have no real explanations. Things like forces, fields, energy, waves, even the elemental particles. We treat these phenomena as fundamental, the irreducible building blocks of our vision of reality. This process is not new. It has been at the heart of science since the beginning and continues to be.

"There had been an attempt to explain electromagnetic phenomena in terms of physical laws that were already understood, involving mechanical principles and the like, but this was unsuccessful. It turned out that to explain electromagnetic phenomena, features such as **electromagnetic charge and electromagnetic forces had to be taken as fundamental**, and [in the nineteenth century] Maxwell introduced new fundamental electromagnetic laws."

Id., p.127 (emphasis added).

"Th[e] view of space as a sort of primitive container of matter is intuitive, but physical theories increasingly suggest that it's incorrect. Relativity theory suggests that space is not absolute. Many physicists entertain theories in which space isn't present at the fundamental level but emerges only at a higher level." Chalmers, *The Conscious Mind*, p.178.

"In physics, the basic level of reality consists of evanescent quantities, such as quantum wave functions, with no special properties of solidity. Science also tells us that solid objects are mostly empty space. What makes objects count as solid is the way they interact with one another. A solid object is (roughly) one that other objects cannot easily penetrate. Solidity is really defined in terms of a certain pattern of interaction."

Id., pp. 177-178.

The bottom layer keeps changing, but we continue to hit the botton, below which we cannot go. Until the next major discovery.

During the twentieth century, rather dramatic progress was made in explaining the "how" of the somethings we see: general relativity, the hot Big Bang, inflation, clustering, the creation of the heavy elements and so on, leading to The Milky Way, Earth and us. The story is well known, but the narratives of these discoveries, including my own, are quite misleading. We get an image of almost straight-line progress. The real story is much, much more complicated, even somewhat chaotic. The remarkable results achieved can seem almost serendipitous.

Noble laureate astophysicist James Peebles provides an in depth review of the evolution of contemporary cosmology, exploring the contemporaneous disputes and controversies existing at each step along the way. P. J. E. Peebles, *The Whole Truth: A Cosmologist's Reflections on the Search for Objective Reality* (2022). He interestingly describe how some of the hypotheses seem to have gained broad acceptance before the empirical results were obtained, while others remained hotly contested until the evidence became overwhelming.

Peebles claims that most of the theoretical advances began as the product of thought experiments. In an effort to use sociology to understand the history of science, Peebles calls the initial speculative theories "social constructions."

> "Scientists form their own social construction when a theory looks too good to be wrong and enthusiasm runs ahead of empirical evidence. ...Then there are circular constructions, theories that fit given evidence because they were designed to fit the evidence. They also are known as **just so stories... .**"

Peebles, p.50 (emphasis added).

I prefer simply to call them hypotheses. (Peebles notes that "social constructions ... are more charitably termed hypotheses," *id.*, p.193.)

The subsequent efforts to formulate them into falsifiable propositions and then to test them empirically, greatly assisted by ever improving tools and techniques, resulted in the fashioning of the persuasive empirical "constructions" constituting modern cosmology. And, Peebles claims that "[t]he ... empirical results from natural science, presented in the worked example of physical cosmology, add up to a **persuasive case for observer-independent reality.**" *Id.*, p.xii (emphasis added).

I am skeptical.

As observed by Chalmers, the laws of physics are based upon "things" described and defined by numbers. The laws can apply to a world consisting solely of information. Indeed, physics actually tells us nothing about the actual contents of our world.

"Physics tells us nothing about what mass is, or what charge is: it simply tells us the range of different values that these features can take on, and it tells us their effects on other features. As far as physical theories are concerned, specific states of mass or charge might as well be pure information states: all that matters is their location within an information space. This is reflected in the fact that physics makes no commitment about the way these states are realized."

...

"Physics can remain quite neutral on these questions of how its features are realized, and indeed about whether the features are 'realized' in some such way at all. ...As long as these information states have the right relations among them, then everything will be as it needs to be. On this picture of the world, there is nothing more to say. Information is all there is. This is how I understand the 'it from bit' conception of the world. It is a strangely beautiful conception: a picture of the world as pure informational flux, without any further substance to it."

Chalmers, *The Conscious Mind*, pp.302, 303.

"The it-from-bit-from-consciousness view has the advantage of integrating consciousness with structure at a deep level, without trying to reduce consciousness to structure (as materialism does) and without separating consciousness entirely from physical structure (as dualism does)." Chalmers, *Reality+*, p.419.

We feel that there is something more, something physical. Physics does not purport to try to say what that something is. However, physics, unlike mathematics, is expected to match, more or less, our perceptions of our world. So, the laws of physics face constraints beyond those imposed by logic alone, again unlike mathematics.

"There seem to be two main problems with this picture of the world. The first is posed by consciousness itself. It seems that here, we have something over and above a pure information space. Phenomenal properties have an intrinsic nature, one that is not exhausted by their location in an information space, and it seems that a purely informational view of the world leaves no room for these intrinsic qualities.

...

"The second problem is that it is not obvious that the notion of pure informational flux is coherent. One may feel that on this view the world is too lacking in substance to be a world.

...

"One might find it plausible that every concrete difference in the world must be grounded: that is, that it must be a difference in something. ... That view subtracted the world of all intrinsic qualities, leaving a world of causal relations, with nothing, it seemed, to do the causing."

Chalmers, *The Conscious Mind,* p.304.

"The first problem suggests that we have direct knowledge of some intrinsic nature in the world, over and above pure information, in phenomenal properties; and the second suggests that we may need some intrinsic nature in the world, to ground information states." *Id.,* p.304.

Curiously, those additional constraints arise from our perceptions and observations, not from an external reality, at least, not directly. One might speculate that the somethings involved are our experiences of phenomena, that they are phenomenal rather than physical (or informational). Perhaps, it is the phenomenal experiences that are intrinsic to, that constitute, reality.

"[T]he suggestion is that the information spaces required by physics are themselves grounded in phenomenal or protophenomenal properties. ...Every time a feature such as mass and charge is realized, there is an intrinsic property behind it: a phenomenal or protophenomenal property, or a microphenomenal property for short. ...The ultimate differences are these microphenomenal differences.

...

"Physics requires information states but cares only about their relations, not their intrinsic nature; phenomenology requires information states, but cares only about the intrinsic nature."

Chalmers, *The Conscious Mind,* p.305.

"[I]t could simply be a law that when microphenomenal states realize an information state of a certain sort by virtue of the causal relations between them (by the 'difference that makes a difference' principle), then a direct phenomenal realization of the same state will arise [at the macro level]." *Id.,* p.307.

"Structuralism says that theories in physics can be boiled down to their structure—roughly, their mathematical equations and their observational consequences. A theory in physics is true if this structure is really present in the world. If the structure of atomic physics is really present in the world, for example, then atomic physics is true, and atoms exist."

Id., p.175.

Let's turn briefly to quantum mechanics. Chalmers points out that, in addition to displaying unmatched predictive powers, the theory is remarkably simple and elegant.

"Quantum mechanics gives us a remarkably successful calculus for predicting the results of empirical observations, but it is extraordinarily difficult to make sense of the picture of the world that it delivers." Chalmers, *The Conscious Mind,* p.333.

"Most of the substance of quantum mechanics is found in the Schrödinger equation. This is a differential equation that determines how the wave function of a system evolves under almost all circumstances. ... The most important feature here is that it is a linear " *Id.*, p.336.

"The framework of quantum mechanics is so simple and elegant that a basic theory that does not replicate that simplicity and elegance can never be satisfying or fully plausible. ...[T]he framework is so robust that it seems.extraordinary that we should need to postulate a complex apparatus to explain its simple predictions." *Id.*, p.345.

It consists of Schrödinger's wave equation, which reflects the position and pathway of every particle, and the proviso that upon mesurement, the wave "collapses," resulting in a fixed, determined value for the property that was observed.

"Within a classical framework, the state of a physical system can be expressed in very simple terms. The state of a particle, for example, is expressed by giving determinate values for each of a number of properties, such as position and momentum. We can call this sort of simple value a basic value. Within the quantum framework, things are not so simple. In general, the state of a system must be expressed as a wave function, or a state vector. Here, the relevant properties cannot be expressed in simple values, but instead must be expressed as a kind of combination of basic values." *Id.*, p.334.

"This vector is best regarded as a wave, with different amplitudes at different locations in space; the function that takes a location to the corresponding amplitude is the wave function." *Id.*, p.335.

"[Upon collapse t]he resulting state still corresponds to a wave function, but it is a wave function in which all the amplitude is concentrated at a definite position; the amplitude everywhere else is zero. ...The dynamics of collapse are probabilistic rather than deterministic." *Id.*, p.337.

"To predict the results of an experiment, we express the state of a system as a wave function, and calculate how the wave function evolves over time according to the Schrödinger equation, until the point where a measurement is made. Where a measurement is made, we use the amplitudes of the calculated wave function to determine the probability that various collapsed states will result, and to calculate the probability that the measurement will yield any given quantity."

Id., p. 337.

"All this is counterintuitive, but it is not yet paradoxical. If we take this formalism at face value as a description of reality, it is not too hard to make sense of."

Id., p.335.

Yet, what about the strange consequences described by the theory? Are they not "hard to make sense of?" And, of course, we experience a macro world without superposition. What is the relationship between the quantum micro world and our world? "If we take the primacy of the Schrödinger equation seriously, the central question is why, given that the physical structure of the world is like this, do we experience it like this?" *Id.*, p.349. It is one thing to say that the act of measurement establishes a particular position or momentum of a particle, on a probabilistic basis, for the moment of measurement and something quite different to explain a classical-appearing Universe almost 14 billion years old.

There are three serious questions, which may be seen to present fundamental flaws in the theory. First, what constitutes a relevant measurement? By what or by whom, and performed how? Second, what happens after the measurement? Does the particle eventually resume its superposition? Third, what happens, if anything, to the rest of the Universe when a particular particle is subject to measurement?

"According to this postulate, a collapse occurs when a measurement is made, but what counts as a measurement? How does nature know when a measurement is made? 'Measurement' is surely not a basic term in the laws of nature; if the measurement postulate is to be remotely plausible as a fundamental law, the notion of measurement must be replaced by something clearer and more basic."

Id., p.338.

"In this picture, any macroscopic system will usually be in a large-scale superposition if there is no consciousness in the vicinity. Before consciousness evolved, the entire universe was in a giant superposition, until presumably the first speck of consciousness caused its state to suddenly collapse."
Id., p.340.

"The only remotely tenable criterion that has been proposed is that a measurement takes place when a quantum system affects some being's consciousness. ...[T]his criterion is at least determinate and nonarbitrary. The corresponding interpretation of the calculus is reasonably elegant and simple in its form, and it is the only literal interpretation of the calculus that has any wide currency. ...Note that this interpretation presupposes mind–body dualism."

Id., p.339.

Perhaps, the act of measurement has no actual effect on the particle (*i.e.*, on the wave function) but only on our perception of it. Perhaps, there are two separate realities—the one of the physicists' theories and the one of our perceptions (or of the perceptions of any conscious being), the connection being described by quantum mechanics.

"Our minds are part of reality, but there's a great deal of reality outside our minds. Reality contains our world and it may contain many others. We can build new worlds and new parts of reality. We know a little about reality, and we can try to know more. There may be parts of it that we can never know. Most importantly: Reality exists, independently of us. The truth matters. There are truths about reality, and we can try to find them. Even in an age of multiple realities, I still believe in objective reality."

Chalmers, *Reality+*, p.xxiv.

"We cannot see ourselves clearly because we have not built a theory of physics yet that treats observers as inside the universe they are describing: that understanding is muddled across seemingly disparate concepts we refer to as 'matter,' 'information,' 'causation,' 'computation,' 'complexity,' and 'life.'" Walker, p.238.

"[D]espite our natural confidence in our own existence, some scientists challenge it and argue that life may be just an illusion or epiphenomenon, explainable by known physics and chemistry." Walker, p.2.

I am simply trying to suggest how complicated reality is when one includes the presence of an observer and is trying to identify reality from the observer's perspective.

Perception

I have discussed in several places the incredible fact that our sensory perceptions are derivative and formulated in our minds. I quote a detailed description of the physical process of sight above (*supra*, pp.97-8). I discussed the mental functions in *Important Things We Don't Know*, pp.21-35. It is claimed that the only things that we can and do experience directly, unfiltered and unprocessed, are the fruits of our consciousness, our subjective experiences.

"Knowledge of conscious experience is in many important respects quite different from knowledge in other domains. Our knowledge of conscious experience does not consist in a causal relationship to experience, but in another sort of relationship entirely."

"[O]ur access to objects in the environment is mediated, by some sort of causal chain or reliable mechanism. This sort of mediation is appropriate when there is a gap between our core epistemic situation and the phenomena in question, as in the case of the external world: we are connected to objects in the environment from a distance. But intuitively, **our access to consciousness is not mediated at all**. Conscious experience lies at the center... ."

Chalmers, *The Conscious Mind,* pp.193-194, 196 (emphasis added).

I wonder. Unless (until) we have a satisfactory theory of consciousness, I do not see how we can claim that conscious experience is direct or "unmediated." There could be another step between the stimulus and the feeling. I would think so, actually.

So, the theories of particle physics represent our current idea of reality? I have previously commented on the idea of transporting people and things as in Star Trek.

"Like I did, Carroll [Sean M. Carroll, *The Big Picture: On the Origins of Life, Meaning, and the Universe Itself* (2016)] uses Star Trek's famous transporter capability as an example (*see* 'Beam me up, Scotty' in my chapter on Consciousness, *Important Things,* p.471). Carroll concludes that, as long as we do not get multiple Captain Kirks, it does not much matter whether the transporter reuses the original atoms to reconstruct Kirk or utilizes new ones. That makes sense. 'If there were just a single copy, most of us would have no trouble accepting them as the original person. (Using different atoms doesn't really matter; in actual human bodies, our atoms are lost and replaced all the time.)' *Id.,* p.15."

Imaginings, pp.117-8.

David Chalmers uses the example of a statue.

> "A statue is not exactly the same thing as a structure of atoms. Atoms can come and go, and the statue will remain. The statue can be destroyed, but the atoms will remain. The statue depends on human interpretation to make it a statue, but the atoms do not. ...In some cases, human minds may play a role in making an object what it is."

Chalmers, *The Conscious Mind*, p.195.

"You are not made of the same atoms you were a decade ago, but you are the same person with memories from your past self. You are rebuilding yourself across time... ." Walker, p.138.

Another example appears in the coverage of the 2024 Olympics in Paris. "When the Eiffel Tower underwent renovations in the 20th century, they preserved pieces of the original iron and kept them in storage. Those chunks make up the hexagon figure in the middle of the Olympic gold medal." Lukas Weese, "What are 2024 Olympic gold medals made of? Explaining the Eiffel Tower connection," *The Athletic,* July 27, 2024.

In all cases—Captain Kirk, Michelangelo's David and the Effiel Tower—, the object in question is iconic, unmistakable, and recognizable by billions of people. In each case, the constituent matter making up the object changes. Is the object the same, despite its changing composition? In the most important respects, clearly yes. The reason is that the essence of each object is not the elements of which it is made, but the arrangement of those elements—the design. From where does that emerge? Where does it exist? Anywhere other than in the minds of human beings?

In two cases, the objects were intentionally constructed pursuant to the designs, whereas Captain Kirk was the result of natural processes pursuant to a design somehow embedded in the initial embryonic cell. In both situations, the design is a fundamental reality but is utterly irrelevant to and independent of the quarks and forces of which the object is made. Do we need to fall back on emergence?

Technology

"Life exists in biology and technology as far as we know, but we do not understand the transition or continuation of life from biology into technology, any more than we understand the continuation of life when chemistry transitions to biology."

Sara Imari Walker, *Life as No One Knows It: The Physics of Life's Emergence* (2024), p.239.

"In" technology? As in, some machines are alive?

I am not sure what Walker means when she seems to state that the transition from biology to technology has occurred. I think that the issue is whether technology reflected in artificial intelligence will at some point become alive and how we will know it if it happens. There are people who believe that AI devices will become conscious. Will that development severe the connection we have believed to exist between life and consciousness (that only things that are alive can be conscious, even though not all life may be) or will we conclude that such devices are alive?

What about intuition, hunches and insight (or inspiration)? These are important sources of problem solving and creativity. Can we really expect AI to achieve these human capabilities?

There certainly are practical issues limiting the use of AI, such as energy consumption.

> "[T]here are massive differences between biological brains and deep neural networks when it comes to energy efficiency. ...[T]he company Hugging Face ... calculated that one of its models (a 175-billion-parameter network named BLOOM), during an eighteen-day period, consumed, on average, about 1,664 watts. Compare that to the 20 to 50 watts our brains use, despite our having about 86 billion neurons and about 100 trillion connections... ."

Anil Ananthaswamy, *Why Machines Learn: The Elegant Math Behind Modern AI* (2024), p.429.

Undoubtedly, contemporary life in the form of humans has made technology an extension of living beings. But, that does not make the technology alive itself. "You have no doubt experienced, at least in a small way, the sense that instruments are part of your own body. Users of microscopes and telescopes, operators of cranes and mechanical arms, even automobile drivers all have, from time to time, felt a sense of identity with their machines." Bray, *Wetware*, p.47.

"We are using algorithms to interpret data and 'see' the world for us. To understand data from our telescopes we must write algorithms to process them. To interpret data from many of our microscopes—things like the Large Hadron Collider, which collects terabytes of data on particle interactions—we must do the same. This is not unlike how the brain had to co-evolve with the multicellular eye to process the frequency of light data that the eye receives in order for our mammalian brains to construct the images we 'see.'"

Walker, p.234.

But, as Bray stressed: "Above all, they lack something possessed by even the simplest organism: the capacity of independent survival in the real world. Bray, *Wetware*, p.207.

> "[I]nanimate structures do not grow and repair them-selves. They do not adapt to changing circumstances in the same way as the natural structures. Any attempt to match organisms in this regard would call for machines able to mine ores, extract metals and silicon, refine, mill, draw parts, have the equivalent of a factory assembly line plus a hi-tech facility for the production and assembly of microchips... .".

Id.

There may be a good argument that the experiencing of Darwinian evolution is a defining characteristic of life. That, however, is not a very useful criterion for the identification of new life forms, since its occurrence will not be observable within available time spans. But, signs of evolution, as Walker describes, could be practical evidence. Yet, AI .cannot pass that test. Technology is not subject to natural selection—it is the result of intentional design.

A Simulation?

> "I think:
> Simulations are not illusions.
> Virtual worlds are real.
> Virtual objects really exist."
>
> Chalmers, *Reality+*, p.12.

"[T]here is Descartes's problem about the existence of the external world. It is compatible with our experiential evidence that the world we think we are seeing does not exist; perhaps we are hallucinating, or we are brains in vats." Chalmers, *Reality+,* p.75. Indeed, today, our scientific perception of reality resembles a simulation—a few stipulated rules (Natural Laws) and a bit of "fluctuations" (quantum) to turn it on, and here we are.

Chalmers argues that a computer simulation creating a virtual reality is real. It exists, just made of "bits." That seems right. More controversially, he concludes that the virtual reality thereby created is also real and is or becomes part of the existing reality. This step is a bit more of a stretch because the virtual reality is arguably there only for the observers or participants.

He claims that there could be no way to distinguish a perfect simulation from actual reality. Okay, but what are the odds of a perfect simulation? Simulations containing shortcuts, simplifications and mistakes would seem to be much more likely, based on human experience. (Of course, a superhuman entity might have a better track record.)

Chalmers also speculates that in the future there may be thousands of new simulations created, and he claims that a person then existing would statistically be much more likely to be living in one of the thousands of simulations than in the one real world (assuming one exists). I find this conclusion quite suspect. If, in that future, a "person" suddenly appeared, it may be more likely that he or she was in a simulation if the odds of appearing in any of the then existing worlds were similar. But, why would that be so? Indeed, if that person descended from inhabitants of today's world, then they would almost certainly find themselves in that world rather than any of the new (simulated) worlds.

The point that caught my attention, however, is Chalmers' comment that we have accepted something similarly weird with the recognition by science of quantum mechanics. Our contemporary view of reality is of something we can perceive and measure but that consists of quantum fluctuations, waves, fields, quarks, charges, and forces and contains dark matter and dark energy.

"Science has taught us that there's much more to reality than initially seems to be the case. For millennia, we didn't know that cats and dogs and trees are made of cells, let alone that the cells are made of atoms or that those are fundamentally quantum mechanical. Yet these discoveries about the nature of cats and dogs and trees have not undermined their reality."

...

"[T]rees and flowers certainly don't seem to be digital objects. But they don't seem to be quantum mechanical objects either. Yet deep down they are, and few people think that the mere fact that trees are grounded in quantum processes makes them less real. I think that being digital is just like being quantum mechanical here."

Id., p.118.

"You could worry about the stuff that photons and quarks are made of. Quarks have a special 'quark-ish' intrinsic nature, and a simulation of a quark won't have genuine quarkishness. There's nothing about this sort of quarkishness in modern physics. Quarks are characterized in mathematical terms, and that's that."

Id., p.178.

So,

Does simulated reality really seem that much stranger?

Objects in the simulation can cause other simulated objects to do things. Simulated objects can be have the characteristics of a solid relative to other simulated objects. And, what about us? If our world is a simulation, must we too be simulated? Or, could we be real with all of our experiences, by which I mean all of our sensory inputs, be simulated? Presumably, if we are real, then our experience of things would be real too.

Indeed, a simulation is the simplest explanation for the world we experience. Simple. Very simple. And, simple is good. Except...

Except that if we live in a simulation, there must be a creator of the simulation. It need not be divine, but it must exist (or have existed). Although, some have argued that God is the simplest explanation for everything; I think that that explanation is not so simple.

"Again, there is no fundamental difference between the idea that our universe is a simulation created by hyperadvanced intelligent beings and the idea that it was created by a god with unimaginable powers. ...If our universe is the only universe in reality, it would be extremely improbable to find ourselves in a fine-tuned cosmos by chance. The strong anthropic principle implies design, and it therefore attributes the fine-tuning of the laws and constants to some kind of intelligent creator or programmer."

Seyed B. Azarian, *The Romance of Reality: How the Universe Organizes Itself to Create Life, Consciousness, and Cosmic Complexity* (2024), p.261.

"For us to be inside a simulation requires a programmer external to our universe and a code or program that when executed runs our universe. This is consistent with intelligent design, which would posit a very similar explanation for reality by replacing 'programmer' with 'designer': a designer of the universe exists (perhaps a God) who specified the initial conditions from which our universe was born."

Walker, p.106.

CHAPTER 10

Inevitability

"Was it too bold a thought that among ...
those myriads and milky ways of solar systems
which constituted matter—
one or other of these inner-worldly heavenly bodies
might find itself in a condition corresponding to that
which made it possible for our earth
to become the abode of life?"

Thomas Mann
The Magic Mountain
(1924)
(Kindle Edition, Loc.5468).

I want to add a few comments about inevitability, discussed at various points in *Important Things We Don't Know: About Nearly Everything*. The comments concern three questions. Is the existence of the Universe contingent or inevitable? Is the existence of life contingent or inevitable? And, are our scientific theories contingent or inevitable? (If you prefer, these questions could be framed as "highly unlikely or highly likely," leaving room for "not surprising" as the answer.)

I.

> "The lesson is clear: quantum gravity not only appears
> to allow universes to be created from nothing
> —meaning, in this case, ... the absence of space and time—
> it may require them. 'Nothing'
> —in this case no space, no time, no anything!—
> is unstable."
>
> Lawrence Krauss
> *A Universe from Nothing:*
> *Why There Is Something Rather than Nothing*
> (2012), p.170.

Assuming the same initial state (which we cannot explain), the most interesting question is whether the something like the somethings with which we are familiar was inevitable or contingent. We can start with the recognized Natural Laws. We have no basis for asserting that they were more than a mere happenstance. The same for the Constants, even more so. Thus, the "how" that we think modern cosmology has described cannot have been inevitable. Indeed, it seems highly unlikely.

However, Krauss argues that some type of "something" is inevitable. He reasons that quantum fluctuations will necessarily occur, even in nothingness; that bubbles of matter and antimatter can arise with no net energy and that occasionally inflation will occur creating from some such bubbles universes. Most such universes would presumably be quite unlike ours.

"In quantum gravity, universes can, and indeed always will, spontaneously appear from nothing. Such universes need not be empty, but can have matter and radiation in them, as long as the total energy, including the negative energy associated with gravity, is zero. " Krauss, pp.167-8.

"[I]f inflation indeed is responsible for all the small fluctuations in the density of matter and radiation that would later result in the gravitational collapse of matter into galaxies and stars and planets and people, then it can be truly said that we all are here today because of quantum fluctuations in what is essentially nothing." *Id.*, p.98.

"[T]he possibility that positive energy stuff, like matter and radiation, can be complemented by negative energy configurations that just balance the energy of the created positive energy stuff. In so doing, gravity can start out with an empty universe—and end up with a filled one." *Id.*, p.99.

"[I]f our universe arose from nothing, a flat universe, one with zero total Newtonian gravitational energy of every object, is precisely what we should expect. ...This is the simplest version of nothing, namely empty space." *Id.*, p.149.

But, there is more. With enough such universes (infinitely many), some will be just like ours. (We know that a universe like ours is possible, since ours exists.) And, the anthropic argument emerges, observing that it should be no surprise that we are here, since if we are existing, it has to be in a universe like ours. Note that Krauss acknowledges that the utility of the anthropic argument requires the existence of a multiverse, as for which we have no evidence and little prospect of obtaining any (since the individual universes in the multiverse are totally separate from one another). *See Important Things We Don't Know*, pp.530, 535-59.

'[P]hysicist Steven Weinberg proposed, based on an argument he had developed more than a decade earlier—before the discovery of dark energy—that the 'Coincidence Problem' could therefore be solved if perhaps the value of the cosmological constant that we measure today were somehow 'anthropically' selected. That is, if somehow there were many universes, and in each universe the 5 of the energy of empty space took a randomly chosen value... . ...**This argument, however, makes mathematical sense only if there is a possibility that many different universes have arisen.**"

Krauss, p.125.

"Here arose the first "andscape' in which the anthropic argument ... could play itself out. If there are many different states in which our universe could end up in after inflation, perhaps the one we live in, one in which there is non-zero vacuum energy that is small enough so galaxies could form, is just one of a potentially infinite family and the one that is selected for inquisitive scientists because it supports galaxies, stars, planets, and life." *Id.*, pp.129-30.

"In a multiverse of any of the types that have been discussed, there could be an infinite number of regions, potentially infinitely big or infinitesimally small, in which there is simply 'nothing,' and there could be regions where there is 'something.' In this case, the response to why there is something rather than nothing becomes almost trite: there is something simply because if there were nothing, we wouldn't find ourselves living there!" *Id.*, pp.177-8.

"The possibility that our universe is one of a large, even possibly infinite set of distinct and causally separated universes, in each of which any number of fundamental aspects of physical reality may be different, opens up a vast new possibility for understanding our existence." *Id.*, p.175.

Peebles expresses skepticism about the anthropic argument, asking why our Universe has billions of galaxies rather than just our one. He suggests that perhaps life is so unlikely that it requires that we have billions of possible worlds in order to get one with life.

"This idea is not so easy to assess, to falsify. For example, one might ask why there are so many galaxies in our universe; wouldn't one do? But maybe the formation of something like us is so exceedingly unlikely that a great many galaxies were needed to improve the chance of our existence, which we presume is real." Peebles, pp.142-143. *See also, Id.*, p.203 ("why do we find ourselves in a universe with so many galaxies of stars with planets that would be suitable homes for us? It seems excessive for satisfaction of the anthropic principle").

Yet, "[a] number of central ideas that drive much of the current activity in particle theory today appear to require a multiverse." Krauss, p.126.

Of course, this challenge assumes that the Universe is not filled with life forms of which we simply are unaware. But, if it is not, then we are confronted with the "fine-tuning" problem even with a multiverse.

"What explains the fine-tuning of the multiverse? The multiverse itself is presumably the consequence of some underlying laws or principles. If those laws had been different, there might have been just one universe, not a multiverse. So we need to explain this fine-tuning too. ...We may well have to accept some unexplained impressiveness as a brute fact about the universe. We can try to minimize what needs to be explained, perhaps by boiling it down to simple principles at the fundamental level. But we'll always be left with the problem of why there is something rather than nothing. We'll always be left with the problem of why the ultimate laws are the way they are. And we'll always be left with the problem of why they're so interesting."

Chalmers, *Reality+,* p.133.

II.

"The emergence of cognition, and even intelligence,
was just as inevitable as the emergence of life.
...
"...[A] new scientific and spiritual worldview—
called poetic meta-naturalism...
argues that nature has an intrinsic purpose—
to wake up and experience the fruits of its own creation."
...
"In our quest to understand cosmic evolution, we will
arrive at a 'theory of everything' ... .
...
"This ambitious theory attempts to solve the greatest
remaining mysteries of science. The infamous
'hard problem of consciousness,' the puzzle of free will,
and the mystery of increasing cosmic complexity
in an increasingly entropic universe
all begin to unravel as the unifying theory dissolves the paradoxes
created by the unjustified assumptions of the reductionist worldview... ."

Seyed B. Azarian,
The Romance of Reality:
How the Universe Organizes Itself to Create
Life, Consciousness, and Cosmic Complexity
(2022), pp.3, 73, 255, 6-7, 156.

Really? Let's take a look.

The book is full of repetition. There are lengthy, almost poetic sections on important, but noncontraversial, propositions. I felt like I was just wandering. So, I have tried to extract and organize the key points, then analyze the thesis.

I previously wrote about some theories and research addressing the apparent self-organizing tendencies of inanimate matter. (Perhaps, it is more accurate to say the tendency of certain inanimate matter to become organized, *i.e.*, for certain structures to appear.) I discussed in some detail Adrian Bejan's "constructal law" and Jeremy England's "dissipative adaptation." *Important Things We Don't Know,* pp.569-71, 543-4, 573-4.

"[D]issipative structures [were] described by Ilya Prigogine in work that won him the Nobel Prize in Chemistry in 1977. Ilya discovered how chemical systems that exchange energy and/or matter with their environment exhibit the emergence of different kinds of organization. Examples range from turbulent flows to convection cells to the patterned spots on a leopard. Ilya was a larger-than-life personality and built up an almost cultlike following based on his conviction that dissipative structures offer a general theory to explain much of the natural world, including life ...Dissipative structures include examples where life exploits the fact that open, far-from-equilibrium systems can locally maintain order."

Sara Imari Walker, *Life as No One Knows It: The Physics of Life's Emergence* (2024), pp.19, 20.

"Energy flowing through chemical networks forces intricate patterns of biological design into existence... . ...This law has been formalized to some degree, albeit abstractly, in England's work on dissipative adaptation, and also in Bejan's own work, which applies the constructal law to all kinds of systems—chemical, biological, social, economic, even technological."

Azarian, p.138.

The question is not whether it happens, but why or how? What mechanism or process is at work? Azarian provides a plausible partial answer. In doing so, he undertakes a rather extended (and interesting) discussion of entropy and the Second Law of Thermodynamics. The point, however, can be more simply made: "Nature ...abhors a gradient" (Azarian, p.21) or, as I prefer, a differential. An existing difference in temperature, pressure, concentration, electrical charge, energy or entropy will naturally tend to disappear, to equalize.

> "A gradient is the difference between two interacting systems that creates instability, whether it be a difference in temperature, pressure, chemical concentration, or electrical charge. If such a difference exists, there will be spontaneous flow from one system to the other until that difference, or the gradient, is eliminated, and a stable and inert state of equilibrium is achieved. This happens automatically because nature is simply intolerant of gradients."

Azarian, pp.21-22.

A differential may be stable if a barrier or obstacle separates the two levels, but, if a breach occurs (a hole in the dyke), the differential will disappear as the "matter" from one region "flows" into the other. Azarian argues that such a relationship exists with respect to "free energy" (energy available to do work, like energy flowing from the sun). It will follow pathways to dissipation, generally as heat. If a change in structure occurs by chance that enables the more rapid dissipation of free energy, that new structure will be "favored," that is, will be more stable, more lasting, than the prior structures or arrangements. Thus, in circumstances of disequalibrium, a kind of natural selection will result in disturbances, by trial and error, leading to arrangements that more efficiently dissipate free energy. Those arrangements appear organized.

"How a complex system suddenly transitions into a more organized state is not entirely understood, but it can only occur when the system's components are collectively interacting in such a way that the activity of its parts becomes increasingly coordinated and statistically correlated. ...When the collective effects of these mutually reinforcing interactions reach a critical threshold, in a manner that is purely mechanical but also quite mystical in appearance, global patterns of synchronized activity suddenly emerge, and the system gains some new property or function, appropriately called an emergent property."

...

"Through trial and error, the emergent system inevitably finds arrangements that are better at extracting energy from a fluctuating, statistically noisy stream of energy. Because states that absorb energy are stable, they are selected, and because they create entropy, they are harder to reverse."

Id., pp.53, 104.

Azarian argues that the relentless pressure to dissipate free energy can cause a process of or like metabolism in a more complex structure, leading to life, since living things are efficient converters of free energy into heat.

"With no life-forms to graze on that energy, untapped gradients created a planetary-scale energy imbalance that caused global thermodynamic instability. As a result of this tension, autocatalytic chemical sets resembling primitive metabolism spontaneously emerged to unlock and dissipate free energy excesses at locations where stress was greatest."

Id., p.50.

Thereafter, "because collective molecular arrangements that can extract accessible energy tend to be more complex, progress toward increasingly sophisticated dissipative structures is not improbable—in far-from-equilibrium conditions it may be unavoidable." Id., p.54.

So, "A biosphere will tend to produce an increasingly intelligent species, but the route to general intelligence likely follows a different evolutionary path on each life-producing planet. ...We should expect life to readily emerge on planets with geochemical conditions that are sufficiently similar to those on Earth around 3.8 billion years ago." Id., p.55.

I have discussed the claim of the spontaneous emergence of metabolism in a prior chapter, so let's just assume now that it could have happened. There are then two serious issues concerning what would have happened next.

First, how does the metabolizing structure reproduce? Absent reproduction, this structure, no matter how magnificent, will simply enjoy its individual grandeur in impotent solitude. Second, although Azarian eloquently rhapsodizes about the incredible power of random variation over very long periods of time to explore all avenues for improvement, such exploration must always start from somewhere. And, the then available improvements, given the starting position, may simply be limited. Similarly, some or all of the initial improvements will lead to dead ends. The possibilities may be limitless, it may just not be possible to get there from here.

Azarian's thesis encounters serious tension between two relatively certain facts:

1. "Life appears to have emerged on Earth just about as soon as conditions allowed, which would be a massive coincidence if its emergence was in fact improbable and a product of pure chance." *Id.*, p.15.
2. "[A]biogenesis doesn't appear to have ever been repeated" (*id.*, p.56); it has occurred only once on Earth.

Azarian asserts that the answer probably is that the first life that arose on Earth spread so rapidly and extensively that the available energy was co-opted foreclosing opportunities for other life to emerge. He is confident that if that first life had suffered extinction, new life would have appeared to fill the space.

So, "[a] biosphere will tend to produce an increasingly intelligent species, but the route to general intelligence likely follows a different evolutionary path on each life-producing planet. ...We should expect life to readily emerge on planets with geochemical conditions that are sufficiently similar to those on Earth around 3.8 billion years ago." Id., p.55.

I have previously addressed the issue of whether conscious life was inevitable. *Important Things We Don't Know,* pp.575-82. As support for the hypothesis that life—indeed, intelligent life—is inevitable, Azarian's explanation seems to me to be farfetched and clearly inadequate.

III.

"[W]as the construction of the ΛCDM theory
[*see* below] inevitable? The evidence reviewed here
makes the case that it would happen about as close
to compelling as it can get. And let us note
that this is what we expect ifour present physical cosmology
is a good approximation to objective reality... ."

P. J. E. Peebles
The Whole Truth:
A Cosmologist's Reflections
on the Search for Objective Reality
(2022), pp.192-193.

If we assume that there is an objective reality that exists independent
of any observer and assume that aspects of that reality are accessible to
our senses, would we conclude that the development of science we have
experienced was essentially inevitable? The question goes to the likely
efficacy of our scientific method.

"[W]hat am I to make of the phenomenon, as I think I may put it, of
being conscious and forming opinions about what it all means? Might the
deep complexity of this phenomenon make the idea of objective reality
meaningless? I have nothing more to add to many generations of thought."
Id., p.209.

In simplest terms, the method consists of the formulation of theoretical models, the derivation of the logical deductions from the models
and the effort empirically to test those deductions against observable
data. When some deductions cannot be confirmed by experimental
results of observations, we check and recheck the experiments and the

data, then seek to revise or replace the models to eliminate the falsification. And, the process starts over.

Of course, the formulation of the models does not occur in a vacuum. Some empirical observations are likely to guide the model builder. We should, therefore, be wary of apparent confirmations that were built into the models. Similarly, revisions that simply accommodate the conflicting data need other testing.

> "We might ask instead whether, among the many paths that could have taken us to the ΛCDM cosmology, there are other paths not taken that would have led us to a different empirically satisfactory cosmology. ... I must add that, as a practical matter, the community is not going to turn to the search for another basin of attraction in the phenomenology unless the set of basic assumptions of the ΛCDM theory is falsified."

Peebles, p.192.

"[T]he ΛCDM theory of the large-scale nature of the universe. The CDM part stands for cold dark matter, a hypothetical substance that interacts little if at all with radiation and the matter we are made of. The symbol Λ represents a hypothetical constant that Einstein introduced, and came to despise." Peebles, p.133.

I distinguish between scientific conclusions that seem likely to be reached if we started over again and those that seem to have depended on some particular creative leap of imagination that might never have occurred or that might have taken a different direction. The distinction may be more pragmatic than logical, but I would categorize as "fact" the following:

- The Universe had a beginning.
- It is 13.6-13.8 billion years old.
- The Universe is expanding.
- The rate of expansion has increased.
- Lightwaves can be bent by gravity.
- The cosmic background radiation is essentially smooth.

"There are three main observational pillars that have led to the empirical validation of the Big Bang, so that, even if Einstein and Lemaître had never lived, the recognition that the universe began in a hot, dense state would have been forced upon us: the observed Hubble expansion; the observation of the cosmic microwave background; and the observed agreement between the abundance of light elements—hydrogen, helium, and lithium—we have measured in the universe with the amounts predicted to have been produced during the first few minutes in the history of the universe."

Krauss, p.108.

"The ... celebrated successes are drawn from carefully controlled experiments or observational situations that can be interpreted in terms of reliable predictions by our fundamental theories. We are basing grand conclusions on special situations." Peebles, p.41.

I put as hypotheses the following:

- Inflation
- Dark matter
- Dark energy
- Einstein's Cosmological Principle.

"[N]atural scientists routinely use what clearly are social constructions. Examples are the general theory of relativity and the big bang theory of the expanding universe in 1960, the hypothetical nonbaryonic dark matter in 1990, and the concept of cosmological inflation now." Peebles, p.208.

"[T]he theory Einstein and Hilbert considered so elegant a century ago, and the scientific community accepted on non-empirical grounds a half century ago, now passes tests of predictions on scales ranging from the laboratory to the solar system and on to the observable universe .... It is remarkable that pure thought sometimes points us in the right direction; here is an example. But of course it is not at all surprising that pure thought more often leads us astray."

Id., pp.84-85.

"The history of thinking about Einstein's cosmological principle, the assumption that the universe is the same everywhere on average, is an example. When Einstein introduced arguments for this picture he may not have known that it was quite contrary to the observations. After publication of the proposal, de Sitter told Einstein about the contrary evidence, but Einstein, and with him the community by and large, continued to accept the assumption."

Id., p.86.

"The enormous expansion during inflation was supposed to have stretched out the primeval, maybe quite chaotic, state of the universe; stretched by such a large factor that there are only tiny variations in the initial conditions across the minuscule bit of the primeval universe that we observe well after inflation. ...[T]he enormous expansion would have stretched out the mean radius of curvature of space sections to some enormous value. That would produce very close to flat space sections across the bit of the universe we can observe. ... The inflation scenario was welcomed as a truly elegant idea... . This was a social construction, supported by little in the way of predictions that were not already being discussed for other reasons."

Id., pp.144-145.

So, I would say that the "facts" are probably relatively inevitable conclusions that we would have reached anyway. The hypotheses, less so—more like speculation.

Assuming we have achieved a relatively accurate account of what happened, is it possible that there could be a significantly different theory that explains the same events? It certainly seems so. I elsewhere discussed the fact of different theoretical constructions reaching identical predictions. *Imaginings*, pp.78-85. Peebles cites the same example of the principle of least action, but concludes that our current theory of cosmology was inevitable, at least given the state of the science in the 1960s.

"The applications of the action principle to electromagnetism and general relativity are of lasting value. The principle also figures in the standard model for particle physics, Feynman's sum over histories formulation of quantum physics, and superstring models for a possible next big advance in fundamental physics."

Peebles, pp.63-64.

"Could a different theory of the structure of matter fit a comparable range of empirical evidence, maybe obtained by tools other than all those that have been used to probe matter and arrive at the standard theory? The same question applies to all our natural sciences, of course. A falsification of the idea is not possible, but we can observe that if two different theories of matter passed all the tests from laboratory and particle accelerator experiments, **the interpretation would require an array of coincidences that seems absurd... ."**

Id., p.52 (emphasis added).

Scientific progress, however, seems inevitable, because:

- "Theorists are inventive; there will be promising-looking ideas. Observers will be occupied checking them and searching for hints to where to go next." *Id.*, p.207.

And,

- "Good evidence is even more influential than good ideas... ." *Id.*, p.148.

REFERENCES

Books

Ananthaswamy, A *Why Machines Learn: The Elegant Math Behind Modern AI*, 2024,

Azarian, S.B. *The Romance of Reality: How the Universe Organizes Itself to Create Life, Consciousness, and Cosmic Complexity*, 2024.

Beerbower, J.E. *Important Things We Don"t Know: About Nearly Everything*, 2016/2022.

Beerbower, J.E. *Imaginings: An Addendum to Important Things*, 2024.

Bray, D. *Wetware: A Computer in Every Living Cell*, 2009.

Carroll, S.M. *The Big Picture: On the Origins of Life, Meaning, and the Universe Itself*, 2016.

Chalmers, D.J. *The Conscious Mind: In Search of a Fundamental Theory*, 1996.

Chalmers, D.J. *Reality+: Virtual Worlds and the Problems of Philosophy*, 2022.

Humphrey, N. *Sentience: The Invention of Consciousness*, 2023.

Koch, C. *Then I Am Myself the World: What Consciousness Is and How to Expand It*, 2024.

Kolakowski, L. *Why Is There Something Rather Than Nothing?: 23 Questions from Great Philosophers*, 2007.

Krauss, L. *A Universe from Nothing: Why There is Something Rather Than Nothing*, 2012.

Lane, N. *Vital Question: Energy, Evolution, and the Origins of Complex Life,* 2015.

Lane, N. *Transformer: The Deep Chemistry of Life and Death,* 2022.

Lightman, A. *Probable Impossibilities: Musings on Beginnings and Endings,* 2021.

Martenez Arias, A.,*The Master Builder: How the New Science of the Cell Is Rewriting the Story of Life,* 2023.

Miller, Jr., W.B., *Bioverse: How the Cellular World Contains the Secrets to Life's Biggest Questions,* 2022.

Mukherjee, S. *The Song of the Cell: An Exploration of Medicine and the New Human,* 2022.

Peebles, P. J. E. *The Whole Truth: A Cosmologist's Reflections on the Search for Objective Reality,* 2022.

Pinker, S. *The Blank Slate: The Modern Denial of Human Nature,* 2002, 2016.

Sapolsky, R. *Determined: A Science of Life without Free Will,* 2023.

Tononi, G. *Phi: A Voyage from the Brain to the Soul,* 2012.

Walker, S. I. *Life as No One Knows It: The Physics of Life's Emergence,* 2024.

Articles

Baisas, W L. "For the first time in one billion years, two lifeforms truly merged into one organism," *Popular Science,* April 18, 2024.

Fedorenko, E., Piantadosi, S.T., Gibson, E.A.F. "Language is primarily a tool for communication rather than thought," *Nature* 630, 575–586, 19 June 2024.

Sorensen, R. "Nothingness," *The Stanford Encyclopedia of Philosophy,* Spring 2023 Edition.

Wagh, M. "Every Single Cell in Your Body Could Be Conscious, Scientists Say. That Could Rewrite Everything We Know About Human Evolution," *Popular Mechanics,* June 10, 2024.

Weese, L. "What are 2024 Olympic gold medals made of? Explaining the Eiffel Tower connection," *The Athletic,* July 27, 2024.

Zimmer, C. "Do We Need Language to Think? A group of neuroscientists argue that our words are primarily for communicating, not for reasoning," *NYT.com,* June 19, 2024.

Zimmer, C. "A Test for Life Versus Non-Life: In a new book, physicist Sara Walker argues that assembly theory can explain what life is, and even help scientists create new forms of it," *NYT.com,* July 31, 2024.

POSTSCRIPT

I interject some personal comments, because they bear on my substantive conclusions.

Curiously, I seem to have discovered that I am what the philosophers call a "dualist"—a believer in "dualism," that there is a "person" separate from the body.

I say this is curious because my conclusion arises from a late-life journey. Some 13 years ago, I began my quest to understand science by writing about it. I had no objectives beyond exploring certain prejudices I had formed over the prior 40 years (which I lay out in detail in my first book). Some thirteen years, three books and over a thousand written pages (and hundreds of books read) later, I discover that, in the process, I went somewhere—I found something. And, it was not some where or some thing for which I had been searching, at least, not consciously.

And, my personal experience contributed to the evolution of my beliefs. It is not the result of solely intellectual investigations. ALS gave me the opportunity to observe and experience first-hand the discontinuities between the body and the mind. The slow but unmistakable decline in physical functionality made the differentiation between the body, withering away, and the mind, avidly devouring the world, quite acute. One could not help but feel intensely that one was occupying two worlds. As my daughter writes: "ALS robs someone of so many abilities, but doesn't get to take their joys, their fundamental agency, their relevance or their humanity."

So, dualism. What else?

I find nothing to commend the theories that consciousness is everywhere or that there is some "pool" of consciousness from which we all partake or to which we all contribute. So, from where does the mind come (and, where does it go)?

I fall back on emergence—even though that answer is not actually an answer to the question. When a brain reaches a certain level of complexity or sophistication, consciousness and a mind "emerge," that is, appear. From nowhere?

Perhaps.

We have no better answer.

And, how about immortality? From the standpoint of science, I find nothing either way. Curiously, the strongest "argument" against the possibility of immortality seems to be its "smell" of wishful thinking—that is, as something too "good" to be true. Hardly a dispositive argument.

John majored in economics at Amherst College, receiving a BA in 1970. He received his JD from The Harvard Law School in 1973. Following law school, he did post-graduate research at Trinity College, University of Cambridge. In late 1974, John began a 37-year career as a commercial litigator with a major law firm in New York City. He retired from the practice of law in 2011, after which he relocated to a small village outside of Cambridge, England. In March 2015, however, John was diagnosed with ALS (motor neurone disease). As a result, he decided to return to the U.S., to live in Old Town Alexandria, Virginia, with his daughter Sarah. His son John Eliot and daughter-in-law Megan, with his two grandchildren Hannah and Jeffrey, live nearby. Confined to a wheelchair since 2018, he has been writing.

Other Books

Important Things We Don't Know (About Nearly Everything)
Wanderings of a Captive Mind (Wanderings Part 1)
The Eyes Have It (Wanderings Part 2)
All that Is Gold (Wanderings Part 3)
Still Wandering: Still Wondering (Wanderings Part 4)
Disappointments: Books Not Written, … (Wanderings Part 5)
On Living While Dying: A Decade with ALS
Me (and Mine)
On Politics, History and Ideology
Imaginings: An Addendum to Important Things

* 9 7 9 8 3 3 0 2 2 6 9 3 1 *